Higher Arithmetic

An Algorithmic Introduction to Number Theory

STUDENT MATHEMATICAL LIBRARY
Volume 45

Higher Arithmetic

An Algorithmic Introduction to Number Theory

Harold M. Edwards

Providence, Rhode Island

2000 *Mathematics Subject Classification.* Primary 11–01.

For additional information and updates on this book, visit
www.ams.org/bookpages/stml-45

Library of Congress Cataloging-in-Publication Data

Edwards, Harold M.
Higher arithmetic : an algorithmic introduction to number theory / Harold M. Edwards.
p. cm. — (Student mathematical library, ISSN 1520-9121 ; v. 45)
Includes bibliographical references and index.
ISBN 978-0-8218-4439-7 (alk. paper)
1. Number theory. I. Title.

QA241.E39 2008
512.7—dc22 2007060578

∞ The paper used in this book is acid-free and falls within the guidelines
established to ensure permanence and durability.
Visit the AMS home page at http://www.ams.org/

10 9 8 7 6 5 4 3 2 1 13 12 11 10 09 08

Contents

Preface

It is widely agreed that Carl Friedrich Gauss's 1801 book *Disquisitiones Arithmeticae* [**G**] was the beginning of modern number theory, the first work on the subject that was systematic and comprehensive rather than a collection of special problems and techniques. The name "number theory" by which the subject is known today was in use at the time—Gauss himself used it (*theoria numerorum*) in Article 56 of the book—but he chose to call it "arithmetic" in his title. He explained in the first paragraph of his Preface that he did not mean arithmetic in the sense of everyday computations with whole numbers but a "higher arithmetic" that comprised "general studies of specific relations among whole numbers."

I too prefer "arithmetic" to "number theory." To me, number theory sounds passive, theoretical, and disconnected from reality. Higher arithmetic sounds active, challenging, and related to everyday reality while aspiring to transcend it.

Although Gauss's explanation of what he means by "higher arithmetic" in his Preface is unclear, a strong indication of what he had in mind comes at the end of his Preface when he mentions the material in his Section 7 on the construction of regular polygons. (In modern terms, Section 7 is the Galois theory of the algebraic equation $x^n - 1 = 0$.) He admits that this material does not truly belong to arithmetic but that "its principles must be drawn from arithmetic."

What he means by arithmetic, I believe, is *exact computation,* close to what Leopold Kronecker later called "general arithmetic."[1]

In 21st century terms, Gauss's subject is "algorithmic mathematics," mathematics in which the emphasis is on algorithms and computations. Instead of set-theoretic abstractions and unrealizable constructions, such mathematics deals with specific operations that arrive at concrete answers. Regardless of what Gauss might have meant by his title *Disquisitiones Arithmeticae,* what I mean by my title *Higher Arithmetic* is an algorithmic approach to the number-theoretic topics in the book, most of which are drawn from Gauss's great work.

Mathematics is about reasoning, both inductive and deductive. Computations are simply very articulate deductive arguments. The best theoretical mathematics is an inductive process by which such arguments are found, organized, motivated, and explained. That is why I think ample computational experience is indispensable to mathematical education.

In teaching the number theory course at New York University several times in recent years, I have found that students enjoy and feel they profit from doing computational assignments. My own experience in reading Gauss has usually been that I don't understand what he is doing until he gives an example, so I try to skip to the example right away. Moreover, on another level, in writing this and previous books, I have often found that creating exercises leads to a clearer understanding of the material and a much improved version of the text that the exercises had been meant to illustrate. (Very often, the greatest enlightenment came when writing *answers* to the exercises. For this reason, among others, answers are given for most of the exercises, beginning on page 179.)

Fortunately, number theory is an ideal subject from the point of view of providing illustrative examples of all orders of difficulty. In this age of computers, students can tackle problems with real computational substance without having to do a lot of tedious work. I

[1]See Essay 1.1 of my book [**E3**]. For the relation of general arithmetic to Galois theory, see Essay 2.1.

have tried to provide at the end of each chapter enough examples and experiments for students to try, but I'm sure that enterprising students and teachers will be able to invent many more.

What began as an experiment in the NYU course turned into a substantial revision of the course. The experiment was to see how much of number theory could be formulated in terms of "numbers" in the most primitive sense—the numbers 0, 1, 2, ... used in counting. To my surprise, I found that not only could I *avoid* negative numbers but that I *didn't miss* them. The simple reason for this is that the basic questions of number theory can be stated in terms of congruences, and subtraction is always possible in congruences without any need for negative numbers. Negative numbers have always led to metaphysical conundrums—why should a negative times a negative be a positive?—which cause confusing distractions right at the outset when the meaning of "number" is being made precise. In this book, the meaning of "number" derives simply from the activity of counting and arithmetic can begin immediately. Kronecker's famous dictum, "God created the whole numbers; all the rest is human work," can be amended to say, "nonnegative whole numbers," which is very likely what Kronecker meant anyway.

A central theme of the book is the problem I denote by the equation $A\square + B = \square$, the problem of finding, for two given numbers A and B, all numbers x for which $Ax^2 + B$ is a square. As Chapter 2 explains, versions of this problem are at least as old as Pythagoras, although two millennia later the *Disquisitiones Arithmeticae* still dealt with it. A simple algorithm for the complete solution is given in Chapter 19.

Work on problems of the form $A\square + B = \square$ led Leonhard Euler to the discovery of what I call "Euler's law," the statement that the answer to the question "Is A a square mod p?" for a prime number p depends only on the value of p mod $4A$. This statement, of which the law of quadratic reciprocity is a byproduct, is completely proved in Chapter 29.

When Ernst Eduard Kummer first introduced his theory of "ideal complex numbers" in 1846, 45 years after the publication of *Disquisitiones Arithmeticae,* Gauss said that he had worked out something resembling Kummer's theory for his "private use" when he was writing about the composition of binary quadratic forms in Section 5 of *Disquisitiones Arithmeticae,* but that he left it out of the book because he had not been able to put it on firm ground.[2] Although the proof of quadratic reciprocity given in this book was originally inspired by Gauss's proof using the composition of forms, it is stated in terms closer to Kummer's ideal numbers. Specifically:

If, in addition to using ordinary numbers 0, 1, 2, ... , one computes with a symbol $\sqrt{A}$ whose square is a fixed number A, one has an arithmetic—I have dubbed it the arithmetic of "hypernumbers" for that A—in which the natural generalization of doing computations mod n for some number n is to do computations mod $[a, b]$ for some *pair* of hypernumbers a and b. (With ordinary numbers, the Euclidean algorithm serves to reduce the number of numbers in a set that describes a modulus to just one, but with hypernumbers two may be needed, as is shown in Chapter 18.) With natural definitions of multiplication and equivalence of such "modules of hypernumbers," the computations needed to solve $A\square + B = \square$ and to prove quadratic reciprocity can be explained very simply. In this way, Gauss's difficult composition of forms is avoided but the essence of his method is preserved.

The last two chapters relate the methods of the book to Gauss's binary quadratic forms so students interested in reading further in the *Disquisitiones Arithmeticae*—or students interested in binary quadratic forms—will be able to make the transition.

Finally, an appendix gives a table of the cycles of stable modules of hypernumbers for all numbers $A \leq 111$ that are not squares, which will be useful for students, as they were for me, in understanding the general theory and in working out examples.

---

[2]See [**E4**].

Chapter 1

Numbers

This book deals with numbers of the simplest kind, the ones we learn as children when we learn to *count,* the numbers 0, 1, 2, ... (Zero is included because the outcome of a count can be "none.") They are often called "natural numbers" or "counting numbers" or "nonnegative integers." Here they—and only they—will be called numbers.

Numbers are *ordered* in the sense that two given numbers a and b satisfy either $a < b$ or $a = b$ or $a > b$, meaning that if two counts are done simultaneously, one to a and one to b, either the count to a will finish first, or they will finish at the same time, or the count to b will finish first. Normally numbers are visualized as a *sequence* written from left to right, starting with 0 and listing the numbers in order, continuing (in the imagination) forever. In terms of this image, the order relation becomes the relation of a lying to the left of, or coinciding with, or lying to the right of, b.

Addition of numbers is very close conceptually to the basic meaning of numbers as the outcomes of counts. If a set containing a objects is united with a set containing b objects, the new set will contain $a + b$ objects. The basic properties of addition are commutativity—the statement that $a + b = b + a$—and associativity—the statement that $(a + b) + c = a + (b + c)$. Both of them follow from the very meaning of the operation of counting.

The notions of counting and recording numbers go back to the earliest periods of human prehistory, but our decimal system of writing numbers and computing with them on paper is of comparatively recent origin—in Europe, at any rate, it was still a novelty five hundred years ago—and effective machine computations began to be done only in the 20th century.

The decimal system writes numbers using just the ten symbols 0, 1, 2, 3, 4, 5, 6, 7, 8, 9 for the first ten numbers and describes larger numbers using a place system to represent *powers of ten,* so that 12340 means $1 \cdot 10^4 + 2 \cdot 10^3 + 3 \cdot 10^2 + 4 \cdot 10$. Elementary and familiar as this system is, its power and simplicity are definitely worthy of some attention at the outset of a course in number theory. For example, this system makes it possible for schoolchildren to learn to perform an addition like $12340 + 567890 = 580230$ with little difficulty, a task that five hundred years ago required a skilled professional.

Multiplication of numbers is a much more sophisticated operation than addition, and it is harder to teach to schoolchildren. If a and b are numbers, their *product* is the number ab of objects in a rectangular array of objects that contains a rows and b columns. Since counting ab objects is the same as adding b to itself a times, the problem of computing the product of a and b—the problem of computing ab—can be reduced to addition by the algorithm:

```
Input: Two numbers a and b.
Algorithm:
    Let p = 0 and t = a.
    While t > 0
      Reduce t by 1 and add b to p
    End
Output: p
```

The "while" loop is executed a times in the course of reducing t from a to 0 (if $a = 0$, the loop is never executed and p remains at zero) and each execution of the loop adds b to p, so that the final value of p is the product—the number b added to itself a times.

This algorithm is unusable for hand computation if a is at all large. Amazingly, modern computers are so lightning fast that they can multiply numbers with 4 or 5 digits rather quickly in this primitive way, but such a computation is a pointless waste of their power. A more efficient but still elementary multiplication algorithm is:

```
Input: Two numbers a and b.
Algorithm:
    Let p = 0 and t = a.
    While t > 0
      Let k = 1.
      While t ≥ 10k
        Multiply k by 10
      End
      Reduce t by k and add kb to p.
    End
Output: p
```

This algorithm is geared to the decimal system in which multiplication by powers of 10 is easy—just shift the digits the required number of places to the left. Instead of adding b repeatedly to p, this algorithm finds the largest power of 10 that is less than or equal to a, call it $10^e = k$ (which one can do by inspection in the decimal system), adds 10^e times b to p all at once (to find $10^e b$ does not, of course, require multiplication, just writing e zeros after b), and reduces by 10^e the number of times b still needs to be added to p.

This more efficient algorithm is similar to the algorithm that is taught in school, except that it begins with the leftmost digit of a rather than the rightmost, and it does not assume that multiplication by a single-digit number is easy; for example, if $a = 32$, it generates the product as $p = (10 \times b) + (10 \times b) + (10 \times b) + b + b$ instead of as $p = (2 \times b) + (30 \times b)$ the way the usual algorithm does.

Computers represent numbers in the binary system, not the decimal system, so multiplication by 2, rather than multiplication by 10, is the easy operation for them to do, because in binary arithmetic multiplication by 2 is accomplished by putting a zero to the right of the number. In adapting the above multiplication algorithm for use on a computer, therefore, it is natural to change 10 to 2 in the two places where it occurs. Of course an algorithm for multiplication is hard-wired into the circuitry of the computer where the user never needs to be concerned with it, but students of number theory should give thought to what the circuitry is accomplishing.

The other basic laws of arithmetic

$$ab = ba, \quad (ab)c = a(bc), \quad a(b + c) = ab + ac$$

all follow from the meaning of addition and multiplication in terms of counting and will be taken for granted.

With this narrow meaning of "number," subtraction and division are not always possible.

The symbol $b - a$ represents "the number which, when added to a, gives b," and there is obviously no such number when $b < a$. Therefore, this symbol can only be used legitimately (in view of the meaning of "number" here) after $b \geq a$ has been proved. For example, the last algorithm above could have said that t is to be replaced by $t - k$ because it has determined k in such a way that $t \geq k$.

Division requires a similar restriction. The symbol b/a represents "the number which, when multiplied by a, gives b." For randomly chosen a and b there is very rarely any such number. Again, the notation b/a will be used, but only when b has been shown to be a multiple of a.

However, *division with remainder* works in all cases in which a is not 0: Given two numbers a and b with $a \neq 0$, there are numbers q and r for which $b = qa + r$ and $r < b$. Moreover, q and r are determined by a and b by means of the simple algorithm:

Input: Two numbers a and b with $a \neq 0$
Algorithm:
  Let $q = 0$ and $r = b$
  While $r \geq a$
    Reduce r by a and add 1 to q
  End
Output: The quotient q and the remainder r of the division.

or, more efficiently,

Input: Two numbers a and b with $a \neq 0$
Algorithm:
  Let $q = 0$ and $r = b$
  While $r \geq a$
    Set $k = 1$
    While $r \geq 2ka$
      Multiply k by 2
    End

Reduce r by ka and add k to q
End
Output: The quotient q and the remainder r of the division.

(In decimal arithmetic, 10 would replace 2 in the two places it occurs.)

These two algorithms begin with the solution $(q, r) = (0, b)$ of $b = qa + r$; they modify (q, r) at each step in such a way that $b = qa + r$ remains true and r is reduced, and they terminate when $r < a$. (If $a = 0$, either algorithm endlessly subtracts 0 from b.) In the first algorithm, a is repeatedly subtracted from r and 1 is added to q until r is less than a, but in the second these operations are done in batches—subtracting $2^e a$ from r and adding 2^e to q, where e is as large as possible.

Once again, the core idea is that of *counting*. Arithmetic—the operations of addition, subtraction, multiplication, and division with remainder—are mere elaborations.

Exercises for Chapter 1

Study Questions.

1. Think through how you would explain the meaning and truth of the commutative law $ab = ba$ of multiplication to an intelligent eight-year-old.

2. Do the same for the associative law of multiplication $(ab)c = a(bc)$ and the distributive law $a(b + c) = ab + ac$.

3. Many of the computations in later chapters will deal with very large numbers—fifteen or more digits at times. Somewhat surprisingly, it can be tricky to do such computations on easily available computers using easily available software because most computations are done using floating point arithmetic, which limits the accuracy with which large numbers can be represented. A program called UBASIC is available for Windows computers with which the algorithms in the chapter can be programmed easily. Try to download UBASIC to your own computer if you have Windows, or, if not, try to develop some means of implementing these algorithms on your computer. UBASIC is in fact far more than is needed. All that is needed is a convenient

method of carrying out algorithms for high-level counting like the ones in the chapter.

Computations.

4. On a programmable calculator—or on a computer—implement the two multiplication algorithms given in the text. (See Exercise 3. For this exercise, you needn't get into really large numbers, so round-off error should pose no problem and you can use ordinary computations.) On a computer, though perhaps not on a calculator, you might be surprised to find how quickly the first algorithm finds products of numbers with 3, 4, or 5 digits. For larger numbers, the naive first algorithm will become unworkable, but the speeded up second algorithm should do fine.

5. Since the calculator or computer is doing the calculations, there is no point in using decimal instead binary numbers. Try both 10 and 2 in the algorithms and see whether you find any significant difference in their execution times.

6. On paper, multiply 33 and 21 in the usual way. Convert both numbers to binary and multiply them in binary. Convert the binary result to decimal to verify that the answers coincide.

7. Compute $7 \times 98765432109876543 21 \times 12345678909876543 21$.

8. Write an implementation of the algorithm for division with remainder and use it to find the quotient and the remainder when 987654321098765 4321 is divided by 5432109876. (In this case, the quotient is so large that even a very fast computer cannot find it in a reasonable time by successively adding 1. The more efficient method that adds powers of 2 to q must be used.)

Chapter 2

The Problem $A\square + B = \square$

The Pythagoreans—the followers of the semi-mythic Greek thinker and teacher Pythagoras who lived in the 6th century BCE—are believed to have studied the following sequence of ratios: $\frac{1}{1}$, $\frac{3}{2}$, $\frac{7}{5}$, $\frac{17}{12}$, $\frac{41}{29}$, $\frac{99}{70}$, $\frac{239}{169}$, $\frac{577}{408}$, ... The sequence is defined by two properties. First, the denominator of each ratio is the sum of the numerator and denominator of the preceding ratio and, second, the numerator of each ratio is the sum of its denominator and the denominator of the preceding ratio.

What was it about this sequence that interested the Pythagoreans? It is the *best possible description of the square root of* 2 in the following sense. In each ratio, the square of the numerator is very nearly equal to twice the square of the denominator, so that the square of the ratio (which is not a *number*) is very nearly equal to the ratio of 2 to 1. Specifically, $2 \cdot 1^2 - 1 = 1^2$, $2 \cdot 2^2 + 1 = 9 = 3^2$, $2 \cdot 5^2 - 1 = 49 = 7^2$, $2 \cdot 12^2 + 1 = 289 = 17^2$, $2 \cdot 29^2 - 1 = 1641 = 41^2$, $2 \cdot 70^2 + 1 = 9801 = 99^2$, $2 \cdot 169^2 - 1 = 57121 = 239^2$, $2 \cdot 408^2 + 1 = 332929 = 577^2$, ... The pattern is probably clear—double the first square is alternately one more than or one less than the second square. These observations raise many questions. Will the pattern persist? Are there solutions of $y^2 = 2x^2 \pm 1$ that it misses? Where does it come from? Why does it work? And—the question a mathematician is most likely to ask—is there a similar way to solve $y^2 = Ax^2 \pm 1$ for other numbers A?

Some centuries later, about 250 BCE, the great Greek mathematician Archimedes studied the ratio of the circumference of a circle to its diameter, the ratio we call π. He proved that π is less than $3\frac{1}{7}$ but greater than $3\frac{10}{71}$. This great achievement is related to infinitesimal calculus, not number theory. However, in the course of estimating π, Archimedes needed to estimate $\sqrt{3}$, and he asserted,[1] without explanation, that $\frac{265}{153} < \sqrt{3}$ and $\frac{1351}{780} > \sqrt{3}$. With your calculator you can easily check that $3 \cdot 153^2 - 2 = 70225 = 265^2$ and $3 \cdot 780^2 + 1 = 1825201 = 1351^2$, which proves the inequalities Archimedes states and justifies his use of them, but leaves one wondering where he got them and whether there are others like them. In particular, it is natural to wonder about his use of a solution of $3x^2 - 2 = y^2$ for his lower estimate; a solution of $3x^2 - 1 = y^2$ would seem preferable.

There is a reference in Plato's dialogue *Theaetetus* (written a century before Archimedes' estimate of π) to a mathematician who studied the square roots of numbers up to 17, which suggests that the Greek mathematicians—although we usually think of them as being geometers—were interested in at least one problem in number theory, namely, the solutions of $Ax^2 \pm B = y^2$, where $A = 2, 3, \ldots, 17$ and where B is a small number.

We will use the shorthand $A\Box + B = \Box$ for this problem—given two numbers[2] A and B, find squares x^2 and y^2 for which $Ax^2 + B = y^2$. The problem $A\Box - B = \Box$ is *included* in $A\Box + B = \Box$ when one uses the following trick. If $Ax^2 - B = y^2$, then $A^2x^2 - AB = Ay^2$, which is to say $Ay^2 + AB = (Ax)^2$. Thus, the solutions of $Ax^2 - B = y^2$ coincide with the solutions of $Au^2 + AB = v^2$ in which v is divisible by A. (Set $y = u$ and $x = v/A$.)

---

[1] By the way, Archimedes and the other Greek mathematicians did not even have the convenience of modern decimal notation—much less calculators—and the means by which they did their calculations is not exactly known. Most likely, it involved the use of a counting board or some other abacus-like device.

[2] If A is a square, this problem is easy—see the exercises. Therefore, only the case $A \neq \Box$ will be considered.

A primary objective in the chapters ahead is the *complete solution* of $A\square + B = \square$: For given A and B, describe an algorithm that produces *all* squares that solve the problem. (It will be shown, in particular, that the Pythagorean sequence is an algorithm that produces all solutions of $2\square \pm 1 = \square$.) This can be regarded as a vast generalization of the question, "How could Archimedes have found his approximations to $\sqrt{3}$?" Although the historical motive for studying this problem might seem weak to students with no particular interest in the history of mathematics, the fact is that the path to its solution winds through all of the basic topics of elementary number theory and ends with a complete solution of the problem. It is not so much the problem itself as the methods used in its solution that make it worth studying.

Exercises for Chapter 2

Study Questions.

1. There is a very good reason Archimedes did not give a solution of $3x^2 - 1 = y^2$. See if you can find it. [Hint: It has to do with multiples of 3.]

2. See if you can construct a method of generating a sequence of ratios that alternates between solutions of $3x^2 + 1 = y^2$ and of $3x^2 - 2 = y^2$ analogous to the Pythagorean sequence. [Hint: Don't try to use the same formula for all steps.]

3. Proposition 10 of Book 2 of Euclid's *Elements* states, in words rather than algebraic symbols, that $(2x + y)^2 + y^2 = 2x^2 + 2(x + y)^2$. Prove this as an algebraic identity. See if you can use it to derive the Pythagorean sequence of ratios.

4. Try to find close approximations $5x^2 \sim y^2$ and see if you can devise any methods for generating them.

5. For what numbers A can you find solutions of $Ax^2 + 1 = y^2$? For $Ax^2 - 1 = y^2$? [There is no easy answer. Trial-and-error produces meager results as A varies.]

6. Try to devise plausible ways that Archimedes might have arrived at his approximations to $\sqrt{3}$. [Nobody knows how he did.]

7. Give an algorithm for the complete solution $A\square + B = \square$ in the case in which A is a square. [This is a special case of the problem $\square + B = \square$, which can be solved by observing that a difference of two squares has a factorization pq in which $p = x + y$ and $q = x - y$.]

8. In the computation suggested in question 9 below, it is necessary to be able to determine whether a given large number is a square. Here is an algorithm that accomplishes this.

Input: A number N
Algorithm:
  Let $k = 1$, $t = N$
  While $t \geq k$
    Reduce t by k and increase k by 2
  End
Output: If $t = 0$ print "N is a square", else print "N is not a square"

Figure out why it works.

Computations.

9. Devise a computer program that scans for solutions of $A\square + B = \square$. For example, for a given A and for $x = 1, 2, 3, \ldots$ find the square that is nearest to Ax^2 and keep a record of those that are less than 10 away from a square. This computation is a natural accompaniment to question 5 above.

10. How far do you have to carry the scan for solutions of $13\square + 1 = \square$ before you find one? How about $61\square + 1 = \square$? [The second is a trick question, but not in the way you might imagine.]

11. Carry the Pythagorean sequence far enough to find a 10-digit number x for which $2x^2 + 1$ is a square. What is that square?

Chapter 3

Congruences

The history of mathematics shows over and over how important *notation* is. As was mentioned in Chapter 1, the modern decimal system of writing numbers was a great step forward in arithmetic. Around the same time that decimal numbers began to be used, the idea of a symbolic algebra using letters was also developed,[1] a notational innovation that was crucial to the development of all of modern mathematics. Another example of the importance of notation is the Leibniz notation for calculus, which gave great advantages to those who used it over those who used Newton's notation, even though Newton was a more profound mathematician than Leibniz.

In number theory, a notational advance of comparable importance is the following simple notation introduced by Carl Friedrich Gauss in *Disquisitiones Arithmeticae* in 1801. Along with many great advances in mathematical knowledge, the book put forth a new notation for expressing the relation "m and n leave the same remainder when they are divided by a," namely, the notation[2]

$$m \equiv n \bmod a.$$

[1]These two changes may have been related, because when letters were no longer used to denote numbers, they could be used to denote algebraic quantities.

[2]This is not *precisely* Gauss's notation, but it is the modern version of it. He wrote $m \equiv n$ (mod.a) and omitted the parenthetical reference (mod.a) to the modulus when the modulus could be inferred from the context.

In words, this is read, "m is congruent to n modulo a." The sign $\equiv$ with three lines is called the "congruence sign"; as will be seen, its role is very similar to that of the equal sign $=$ with two lines. The word "modulo" or "modulus" is a Latin word related to "measure,"[3] and the congruence $m \equiv n \bmod a$ means something like "when you ignore multiples of a, the numbers m and n are the same." The phrase "modulo a" is usually contracted to "mod a."

(Note that the definition of $m \equiv n \bmod a$ is meaningless in the case $a = 0$. In view of the lemma below, it is natural to define congruence mod 0 as equality. That is, $m \equiv n \bmod 0$ means $m = n$, which means, of course, that congruences mod 0 can be ignored.)

The symbol $\equiv$ by itself has no meaning, because the modulus a must be specified. Similarly, the notation $m \bmod a$ has no meaning, because the modulus a pertains to a *relation* of congruence. A sign $\equiv$ calls for a modulus and conversely.

Congruence mod a is an *equivalence relation* in the sense that it is reflexive (every m satisfies $m \equiv m \bmod a$), symmetric (if $m \equiv n \bmod a$, then $n \equiv m \bmod a$) and transitive (if $l \equiv m \bmod a$ and $m \equiv n \bmod a$, then $l \equiv n \bmod a$), as follows immediately from the definition.

What makes the congruence notation so powerful is that *congruences can be added and multiplied in the same way equations can be added and multiplied,* which means that familiar computational techniques can be used to do computations with new number-theoretic meanings. For example, congruence notation shows easily that Archimedes was right to be satisfied with a solution of $3\square - 2 = \square$, because $3\square - 1 = \square$ *is impossible,* as one can see in the following way. Division of y by 3 leaves a remainder of 0, 1, or 2, so $y \equiv 0$, 1, or $2 \bmod 3$, which implies, by the theorem below, that $y^2 \equiv 0^2$, 1^2, or $2^2 \equiv 1 \bmod 3$. Then, again by the theorem below, $y^2+1 \equiv 0+1 \bmod 3$ or $y^2 \equiv 1 + 1 \bmod 3$. Thus, $y^2 + 1 \equiv 1$ or $y^2 + 1 \equiv 2 \bmod 3$,

[3] In the standard translation of Euclid's *Elements,* the statement that a number "measures" another is synonymous with the statement that the second number is a multiple of the first.

which means that even the congruence $y^2 + 1 \equiv 3x^2 \bmod 3$ is impossible (by the theorem, $3x^2 \equiv 0 \bmod 3$), not to mention the equation $y^2 + 1 = 3x^2$.

Theorem. *If a, k, l, m, and n are numbers for which $k \equiv l \bmod a$ and $m \equiv n \bmod a$, then $k + m \equiv l + n \bmod a$ and $km \equiv ln \bmod a$.*

In short (and loosely stated) when congruent numbers are added or multiplied, the results are congruent, just as equals added to equals or multiplied by equals are equal. The more correct way to state this is to say that *congruence mod a is consistent with addition and multiplication* of numbers.

This theorem is an easy consequence of:

Lemma. *To say $m \equiv n \bmod a$ is the same as to say that there are numbers s and t for which $m + sa = n + ta$.*

Proof of the lemma. Let a, m, and n be given. Say division of m and n by a gives $m = q_m a + r_m$ and $n = q_n a + r_n$, respectively. By definition, $b \equiv c \bmod a$ means $r_n = r_m$, so it implies $m + q_n a = q_m a + r_m + q_n a = q_m a + r_n + q_n a = n + q_m a$, so $s = q_n$ and $t = q_m$ gives an equation of the required form. Conversely, if $m + sa = n + ta$, then $r_m = r_n$ because both are the remainder when $(q_m + s)a + r_m = (q_n + t)a + r_n$ is divided by a. □

Proof of the theorem. The lemma implies that if a, k, l, m, and n are numbers for which $k \equiv l \bmod a$ and $m \equiv n \bmod a$, then there are numbers s, t, u, and v for which $k + sa = l + ta$ and $m + ua = n + va$. Addition of these two equations gives $k + m + (s + u)a = l + n + (t + v)a$ which, by the lemma, implies $k + m \equiv l + n \bmod a$. Similarly, multiplication gives $(k + sa)(m + ua) = (l + ta)(n + va)$ or $km + (ku + sm + sua)a = ln + (lv + tn + tva)a$, so $km \equiv ln \bmod a$, as was to be shown. □

Exercises for Chapter 3

Study Questions.

1. What numbers less than 5 are squares mod 5? (That is, which of the congruences $x^2 \equiv 0 \bmod 5$, $x^2 \equiv 1 \bmod 5$, ..., $x^2 \equiv 4 \bmod 5$ have solutions x?) What numbers less than 14 are squares mod 14? Write an algorithm that determines, for a given modulus a and a given number N, whether N is a square mod a.

2. Given that $x \equiv 2 \bmod 3$ and $x \equiv 1 \bmod 5$, what can you conclude about x?

3. Given that $x \equiv 5 \bmod 6$ and $x \equiv 3 \bmod 7$, what can you conclude about x?

4. Given that $4x \equiv 7 \bmod 15$, what can you conclude about x?

5. Given that $4x \equiv 6 \bmod 14$, what can you conclude about x?

6. Prove that *subtraction modulo a* makes sense (provided $a \neq 0$) in that: for given numbers n, m, and a, the congruence $x + n \equiv m \bmod a$ always has a solution, and any two solutions x are congruent mod a. This observation means that *minus signs can be used with abandon in congruences.* In other words, in a congruence mod a, for any number m, the symbol $-m$ has a meaning, namely, it means any number x that solves the congruence $x + m \equiv 0 \bmod a$.

7. Prove that *division modulo a* does not make sense in an analogous way. In other words, prove it is *not* true that: for given numbers n, m, and a, the congruence $nx \equiv m \bmod a$ has a solution, and any two solutions x are congruent mod a. (Well, obviously the case $n = 0$ must be excluded—division by 0 is meaningless—but adding the assumption $n \neq 0$ or even the assumption $n \not\equiv 0 \bmod a$ is not enough to make the above assertion true.)

8. What, in simple terms, is the meaning of congruence mod 1?

Computations.

9. Write a program that scans for solutions x of the congruence $mx \equiv 1 \bmod a$ for given inputs m and a. What patterns, if any, can you use to distinguish between cases where solutions exist and cases where they do not?

10. Write a program that scans for solutions x of the simultaneous congruences

$$x \equiv m \bmod a$$

and

$$x \equiv n \bmod b$$

for given inputs m, n, a, and b. What patterns, if any, can you use to distinguish between cases where solutions exist and cases where they do not?

Chapter 4

Double Congruences and the Euclidean Algorithm

Writing $m \equiv n \bmod a$ in the form $m + sa = n + ta$ where s and t are numbers (see the lemma of Chapter 3), suggests a type of *double* congruence:

Definition. *Given two nonzero numbers a and b, two other numbers m and n will be said to be congruent* mod $[a, b]$, *written* $m \equiv n \bmod [a, b]$ *if there are numbers s, t, u, and v for which $m + sa + tb = n + ua + vb$.*

Loosely speaking, the relation means that a step from m to n can be accomplished using a combination of steps of size a and steps of size b. (Starting at m, take s steps of size a and t steps of size b to the right, followed by u steps of size a and v steps of size b to the left to end at n.)

For example, the equation $17+100\cdot 60+3\cdot 111 = 5+4\cdot 60+55\cdot 111$ (both numbers are 6350) shows that $17 \equiv 5 \bmod [60, 111]$.

One way of looking at a very important fact of elementary number theory—perhaps the most important fact of elementary number theory—is to say that such a notion of double congruence can in fact be reduced to an ordinary single congruence:

Theorem. *Let a and b be given numbers, neither of them zero.*[1] *There is a third number c with the property that two numbers are congruent* mod $[a, b]$, *as defined above, if and only if they are congruent* mod c. *In fact, c can be found by Euclid's algorithm "subtract the lesser from the greater" as explained below.*

In the case $a = 60$ and $b = 111$ mentioned above, the rule "subtract the lesser from the greater" calls for replacing congruence mod $[60, 111]$ with congruence mod $[60, 51]$ (the lesser 60 is subtracted from the greater 111). Repetition of the algorithm continues: $[9, 51]$, $[9, 42]$, $[9, 33]$, $[9, 24]$, $[9, 15]$, $[9, 6]$, $[3, 6]$, $[3, 3]$. At this point, there is no "lesser" and it is unclear how to proceed, but in fact there is no need to proceed, because congruence mod $[3, 3]$ is the same as congruence mod 3—both just mean that you can go from one number to the other in steps of size 3.

Proof of the theorem. Loosely speaking, the statement to be proved is that if $a < b$ and if you can go from m to n in steps of size a and b, then you can go from m to n in steps of size a and $b - a$, and conversely. This is true, because a step of size $b - a$ can be accomplished by taking a step of size b in one direction and then a step of size a in the opposite direction, while a step of size a can be accomplished by a step of size $a - b$ followed by a step of size b in the same direction.

More precisely, what is to be shown is that if $a < b$ and $m + sa + tb = n + ua + vb$ for some numbers s, t, u, v, then $m + Sa + T(b - a) = n + Ua + V(b - a)$ for some numbers S, T, U, V, and conversely. That the first equation implies the second comes from rewriting the first equation in the form $m + (s + t)a + t(b - a) = n + (u + v)a + v(b - a)$. That the second implies the first follows from adding $(T + V)a$ to both sides to put the second equation in the form $m + (S + V)a + Tb = n + (U + T)a + Vb$.

[1] If either a or b is zero, then the theorem is true in a trivial way because congruence mod $[0, b]$ is the same as congruence mod b and congruence mod $[a, 0]$ is the same as congruence mod a, as one sees directly from the definition.

When $a > b$, an analogous argument applies with the roles of a and b reversed.

Thus, congruence mod $[a, b]$ is the same as congruence mod $[a, b-a]$ when $a < b$ and the same as congruence mod $[a-b, b]$ when $a > b$. Repetition of this process until[2] the two numbers are the same (in the above example they both become 3) proves the theorem, because it shows that congruence mod $[a, b]$ is the same as congruence mod $[c, c]$ for some c, and it is clear from the definition that congruence mod $[c, c]$ is the same as congruence mod c. □

The algorithm can be extended to lists of more than two numbers as well:

Input: A list of nonzero numbers $a_1, a_2, \ldots, a_n$
Algorithm:
  While $n > 1$
    If $a_1 = a_2$ drop a_1 from the list and reduce all subscripts by 1
    Elseif $a_1 < a_2$ subtract a_1 from a_2
    Else subtract a_2 from a_1
  End
Output: The list containing one number with which the algorithm terminates

Clearly this algorithm terminates after a finite number of iterations. (Each iteration reduces the number of numbers in the list when the first alternative holds and otherwise reduces the total of the numbers in the list, so the algorithm must terminate before $n + a_1 + a_2 + \cdots + a_n + 1$ steps have been executed.)

When neither a nor b is zero, the number c that describes the double congruence mod $[a, b]$ is the **greatest common divisor** of a and b because it divides both a and b (both a and b are zero mod $[a, b]$, so both are zero mod c, which is to say that both are divisible by c) and is divisible by any number that divides both a and b (since c is zero mod c, it is zero mod $[a, b]$, which is to say that $c+sa+tb = 0+ua+vb$ for some s, t, u, and v, so if a and b are both zero mod d, so is c).

[2]As long as the numbers are not equal, a repetition of the process reduces the total of the two numbers. Since the reduction of the total cannot continue *ad infinitum*, the repetitions must eventually end, which means that the numbers will have become equal.

More generally, congruence mod $[a_1, a_2, \ldots, a_n]$ is the same as congruence mod c where c, the output of the above algorithm, is the **greatest common divisor** of $a_1, a_2, \ldots, a_n$, because it divides all of the a_i and is divisible by any number that divides all of the a_i. A list $a_1, a_2, \ldots, a_n$ that includes zeros has a greatest common divisor in the same way—simply ignore the zeros, because the greatest common divisor of the nonzero entries in the list divides the zeros as well—unless the list contains *only* zeros, in which case there is no greatest common divisor because all numbers divide all entries.

Two numbers a and b are **relatively prime** if all numbers are congruent to 0 mod $[a, b]$. This is clearly true if and only if at least one of a and b is nonzero and their greatest common divisor is 1.

Exercises for Chapter 4

Study Questions.

1. The example $[60, 111]$ was chosen in an attempt to conceal the greatest common divisor 3. Try to construct other pairs of small numbers that appear at first glance to be relatively prime but in fact are not.

2. Alter the algorithm in the text in such a way that instead of subtracting the lesser from the greater, it subtracts 2^e (or, if you are planning to use the algorithm for pencil-and-paper computation, 10^e) times the lesser from the greater, where e is the largest number for which this subtraction is possible.

3. The equation $\frac{1}{6} + \frac{1}{3} = \frac{1}{2}$ (these are fractions, not numbers!) has meaning mod a whenever a is relatively prime to 6. For example, one can say that $\frac{1}{6} \equiv 2 \bmod 11$ provided one overlooks its evident meaninglessness and sees instead the very clear meaning: "$x = 2$ solves the congruence $6x \equiv 1 \bmod 11$." (On the meaning of $1/6$ see Chapter 1.) For the same reason, one can stretch the notation to say that $\frac{1}{3} \equiv 4 \bmod 11$ and $\frac{1}{2} \equiv 6 \bmod 11$ because $3 \cdot 4 \equiv 1 \bmod 11$ and $2 \cdot 6 \equiv 1 \bmod 11$, respectively. Note that, with these interpretations, $\frac{1}{6} + \frac{1}{3} \equiv \frac{1}{2} \bmod 11$. Work out the analogous congruence for other values of the modulus that are relatively prime to 6 like $a = 19$ or 43 or 143. Why does it always work out this way?

Computations.

4. Write computer implementations of the algorithm of the text and of the speeded up version in question 2 above. You should find that they work very well, even for large inputs a and b, except that the one that is not speeded up can occasionally get stuck doing a lot of subtractions as in the case $a = 1{,}000{,}000{,}000$, $b = 1{,}000{,}000{,}001$.

Chapter 5

The Augmented Euclidean Algorithm

Let a and b be nonzero numbers, and let d be their greatest common divisor. Loosely speaking, d is the size of the smallest step one can take by combining steps of size a and steps of size b. The augmented Euclidean algorithm determines *how* to take a step of size d using steps of size a and steps of size b. More precisely, it determines *two* ways of taking a step of size d using steps of size a and steps of size b; the steps of size a predominate in one and the steps of size b predominate in the other.

In formulas, the algorithm finds solutions u, v, x, y of the pair of equations

$$
\begin{aligned}
d + ub &= va \\
d + xa &= yb
\end{aligned} \tag{1}
$$

where a and b are given nonzero numbers and d is their greatest common divisor. One can take a step of size d to the right either by taking v steps of size a to the right and u steps of size b to the left, or by taking y steps of size b to the right and x steps of size a to the left.

The **augmented Euclidean algorithm** can be formulated in the following way:

Input: Two nonzero numbers a and b
Algorithm:
  Let $d = a$, $e = b$, $u = 0$, $v = 1$, $x = 0$, $y = 1$
  While $d \neq e$
    If $d > e$, change d to $d - e$, u to $u + y$, and v to $v + x$
    Else change e to $e - d$, x to $v + x$, and y to $u + y$
  End
Output: Equations $d + ub = va$ and $d + xa = yb$

If u, v, x, and y are ignored, the algorithm is simply the Euclidean algorithm "subtract the lesser from the greater" and it terminates with both d and e equal to the greatest common divisor of a and b as in Chapter 4. The equations $d + ub = va$ and $e + xa = yb$ hold at the outset because $d + 0 \cdot b = 1 \cdot a$ and $e + 0 \cdot a = 1 \cdot b$ hold at the outset. Each step preserves the truth of these equations, as can be seen in the following way. If $d + ub = va$ and $e + xa = yb$ both hold, their sum $d + (u + y)b = e + (v + x)a$ holds. If $d > e$, the step of the algorithm leaves e, x, and y unchanged, so the equation $e + xa = yb$ remains true, while $d + ub = va$ becomes $(d - e) + (u + y)b = (v + x)a$ which is true because it results when e is subtracted from both sides of the sum. In the same way, if $d < e$, the equation involving d is unchanged and the equation involving e is changed to the equation obtained by subtracting d from both sides of the sum. Thus, both equations are true at each step, including the last step, at which $d = e$, which shows that the output equations are true.

The working of the algorithm can be seen clearly if the steps are shown in tabular form, with one column for each of d, e, u, v, x, and y and with one row for each step of the algorithm.

For example, when $a = 23$ and $b = 14$, the table takes the form

d	e	u	v	x	y
23	14	0	1	0	1
9	14	1	1	0	1
9	5	1	1	1	2
4	5	3	2	1	2
4	1	3	2	3	5
3	1	8	5	3	5
2	1	13	8	3	5
1	1	18	11	3	5

ending with the equations $1 + 18 \cdot 14 = 11 \cdot 23$ and $1 + 3 \cdot 23 = 5 \cdot 14$.

One of the main facts of applied number theory is that *the augmented Euclidean algorithm is quite practical, even when the input numbers are enormous.* Thus, the solutions of (1) for any given pair of nonzero numbers a and b can be found with ease. However, this is true only after the algorithm is modified, as the Euclidean algorithm was modified in Chapter 4, so that it subtracts convenient *multiples* of the lesser from the greater:

Input: Two nonzero numbers a and b
Algorithm:
  Let $d = a$, $e = b$, $u = 0$, $v = 1$, $x = 0$, $y = 1$
  While $d \neq e$
    Let $k = 1$
    While $d > 2ke$ or $e > 2kd$
      Multiply k by 2
    End
    If $d > e$, change d to $d - ke$, u to $u + ky$, and v to $v + kx$
    Else change e to $e - kd$, x to $kv + x$, and y to $ku + y$
  End
Output: Equations $d + ub = va$ and $d + xa = yb$

This "speeded up" version of the basic algorithm simply finds the largest power of 2, call it $k = 2^e$, for which the basic algorithm will repeat the same step k times in a row, and performs these k steps all at once. In the example above, the speeded up algorithm produces the same calculation as the basic algorithm except that, at the end, instead of subtracting 1 from 4 three times in a row, the speeded up algorithm first subtracts it *twice* in a single step and then subtracts it once more. The effect in this case is simply to skip the third line from the bottom in the table, which is scarcely an improvement in the computation. On the other hand, the speeded up algorithm is a huge improvement if at one point of the algorithm a step is encountered in which the smaller of d and e is *much* smaller, as is the case, for example, in the first step of the algorithm when it is applied to $a = 1002$ and $b = 5$.

Exercises for Chapter 5

Study Questions.

1. (a) Find a multiple of 123 that is one more than a multiple of 458, and find a multiple of 458 that is one more than a multiple of 123, showing the working of the algorithm in full.

(b) Find a solution of the simultaneous congruences $x \equiv 100 \bmod 123$ and $x \equiv 300 \bmod 458$.

(c) Find the *smallest* solution of these simultaneous congruences.

2. (a) Show that if a and b are nonzero numbers and d is their greatest common divisor, then a solution (x, y) of $d + xa = yb$ implies a solution in which $x < b$. (b) Show that there is *at most one* solution (x, y) in which $x < b$. [Reduce to the case $d = 1$ and regard the equation as a congruence mod b.]

3. Experience with the augmented Euclidean algorithm leads one to expect that the solutions of equations (1) it produces are the smallest possible ones, which is to say that $x < b$ and $u < a$. That this is indeed the case can be seen by restating the augmented Euclidean algorithm in the following way:

Input: Two nonzero numbers a and b
Algorithm:
  Set $u = x = 0$ and $v = y = 1$
  While $\frac{a}{b} \neq \frac{u+y}{v+x}$
    If $\frac{a}{b} < \frac{u+y}{v+x}$ set $y = u + y$ and $x = v + x$
    Else set $u = u + y$ and $v = v + x$
  End
Output: The equations $d + ub = va$ and $d + xa = yb$ where d is the greatest common divisor of a and b

(In accordance with the definition of "number" in Chapter 1, the fractions $\frac{a}{b}$, $\frac{u+y}{v+x}$ are not numbers. The statements $\frac{a}{b} \neq \frac{u+y}{v+x}$ and $\frac{a}{b} < \frac{u+y}{v+x}$ are shorthand for the statements $a(v+x) \neq b(u+y)$ and $a(v+x) < b(u+y)$.)

(a) Prove that this algorithm produces the same (finite) sequences of values of u, v, x, and y that the augmented Euclidean algorithm does.

(b) Prove that if $xu + 1 = yv$ and if $\frac{p}{q}$ is a fraction satisfying $\frac{u}{v} < \frac{p}{q} < \frac{y}{x}$, then $q > v$ and $q > x$. (This is the fundamental fact in the theory of *Farey series*. See, for example, [**E1**, p. 264].)

(c) Use (b) to show that the final value of x is less than b and the final value of u is less than a.

4. Given just *one* of the equations of (1), there is an easy way to determine the other. Find it.

5. When a is relatively prime to b, the number v in (1) is called a *reciprocal of a mod b*. Explain. Thus, Exercise 3 of Chapter 4 states that if b is relatively prime to 6, then the reciprocal of 2 mod b is the sum of the reciprocal of 6 mod b and the reciprocal of 3 mod b.

6. How can a reciprocal of a mod b be used to solve a congruence of the form $ax \equiv c \bmod b$ for x when a, b and c are nonzero numbers and a and b are relatively prime?

7. Show that, when a and b are relatively prime, *division by a mod b is possible* in the sense that every congruence $ax \equiv c \bmod b$ has a solution x for every c and that any two solutions x are congruent mod b.

8. Exercise 3 implies that if a and b are relatively prime, then the number v, which is the reciprocal of a mod b, determines not only $u = \frac{va-1}{b}$ but also x and y in equations (1). Thus, the entire output of the augmented Euclidean algorithm can be deduced from the solution of the congruence $ax \equiv 1 \bmod b$. This can be accomplished by the following alternative algorithm:

Let the problem be to solve $ax \equiv c \bmod b$ when a, b, and c are given numbers and $a \leq b$. (If $a > b$, use the algorithm that follows to find the reciprocal of b mod a, from which the reciprocal of a mod b

can be deduced.) Let ma be the least multiple of a that is greater than b. The desired x satisfies $max \equiv mc \bmod b$, which is to say $a_1x \equiv c_1 \bmod b$, where $a_1 = ma - b$ and $c_1 = mc$. If a does not divide b, then $a_1 < a$ and the new problem has the same form as the original problem, except that a is reduced. Repeated application of this reduction method must eventually reach a problem of this form in which a divides b. But $ax \equiv c \bmod qa$ has a solution if and only if a divides c, in which case the most general solution is $x \equiv \frac{c}{a} \bmod \frac{b}{a}$. (In particular, if $a = 1$, there is always a solution and the most general solution is $x \equiv c \bmod b$.)

Express this method of solving $ax \equiv c \bmod b$ as a formal algorithm.

Computations.

9. Implement the augmented Euclidean algorithm on a computer and see for yourself how well it works even with numbers that have many, many digits.

10. The number 200560490130 is the product $2 \cdot 3 \cdot 5 \cdot 7 \cdot 11 \cdot \cdots \cdot 29 \cdot 31$ of the first 11 prime numbers. For various large numbers, find their greatest common divisors with this number. What is the largest number m for which it is true that "a number with m digits that is relatively prime to 200560490130 is prime"?

Chapter 6

Simultaneous Congruences

Theorem. *Let a and b be relatively prime nonzero numbers. For any given numbers m and n, the simultaneous congruences $x \equiv m \bmod a$ and $x \equiv n \bmod b$ are equivalent to a single congruence $x \equiv k \bmod ab$ for some k.*

Proof. The assumption that a and b are relatively prime means that their greatest common divisor is 1, so the augmented Euclidean algorithm finds solutions of $1 + ub = va$ and $1 + xa = yb$. Then $va \equiv 0 \bmod a$ and $va \equiv 1 \bmod b$, while $yb \equiv 1 \bmod a$ and $yb \equiv 0 \bmod b$. The simultaneous congruences $x \equiv m \bmod a$ and $x \equiv n \bmod b$ can then be solved simply by setting $x = m \cdot yb + n \cdot va$, because mod a this number is $m \cdot 1 + n \cdot 0 = m$ and mod b it is $m \cdot 0 + n \cdot 1 = n$.

Thus, the congruences have a solution for any given m and n. Since $x \equiv x' \bmod ab$ implies $x \equiv x' \bmod a$ and $x \equiv x' \bmod b$, any number congruent to $myb + nva \bmod ab$ is a solution, so there is a solution less than ab.

It remains to show that there is *only one* solution less than ab. This follows from the observation that there is a solution less than ab of each of the ab problems $x \equiv m \bmod a$ and $x \equiv n \bmod b$ in which $m < a$ and $n < b$, so no two numbers less than ab can solve the same problem. □

This theorem is the case $l = 2$ of:

The Chinese Remainder Theorem[1]. *Let $a_1, a_2, \ldots, a_l$ be nonzero numbers with the property that a_i and a_j are relatively prime whenever $i \neq j$, and let $m_1, m_2, \ldots, m_l$ be given numbers. The simultaneous congruences*

$$x \equiv m_i \bmod a_i \qquad (i = 1, 2, \ldots, l)$$

have a solution, and any two solutions are congruent $\bmod\, a_1 a_2 \cdots a_l$.

Proof. Since a_1 and a_2 are relatively prime, the theorem above shows that there is a number k_1 with the property that the simultaneous congruences $x \equiv m_1 \bmod a_1$ and $x \equiv m_2 \bmod a_2$ are equivalent to the single congruence $x \equiv k_1$. Therefore, the three congruences $x \equiv m_i \bmod a_i$ for $i = 1, 2, 3$ are equivalent to the pair of congruences $x \equiv k_1 \bmod a_1 a_2$ and $x \equiv m_3 \bmod a_3$. The above theorem then shows that these two reduce to a single congruence of the form $x \equiv k_2 \bmod a_1 a_2 a_3$ once it is shown that $a_1 a_2$ and a_3 are relatively prime, which is a consequence of:

Lemma. *If a and b are both relatively prime to c, then ab is relatively prime to c.*

Proof of the lemma. The assumption that a is relatively prime to c implies that there is a solution (r, s) of $1 + rc = sa$. In the same way, there is a solution (t, u) of $1 + tc = ub$. Then the equation $sa \cdot ub = (1 + rc)(1 + tc) = 1 + (r + t + rtc)c$ shows that no number greater than 1 can divide both ab and c. □

Conclusion of the proof of the Chinese remainder theorem. It has now been shown that the first three congruences are equivalent to a single congruence $x \equiv k_2 \bmod a_1 a_2 a_3$ for some k_2. Therefore, the first four are equivalent to just two congruences $x \equiv k_2 \bmod a_1 a_2 a_3$ and $x \equiv m_4 \bmod a_4$, which are in turn equivalent to $x \equiv k_3 \bmod a_1 a_2 a_3 a_4$ for some k_3 because the lemma proves first that a_4 is relatively prime to $a_1 a_2$ and then that it is relatively prime to $a_1 a_2 a_3$.

---

[1] Closely related statements are to be found in ancient Chinese texts.

Continuing in this way, the given set of l congruences reduces to a single congruence $x \equiv k_{l-1} \bmod a_1 a_2 \cdots a_l$ for some k_{l-1}. In particular, any two solutions are congruent mod $a_1 a_2 \cdots a_l$. □

Exercises for Chapter 6

Study Questions.

1. The *Introduction to Arithmetic* of Nicomachus, a Greek work of the 2nd century, poses and solves the problem of finding a number whose remainders when divided by 3, 5, and 7 are 2, 3, and 2, respectively. Find the most general solution.

2. Show explicitly the correspondence the Chinese remainder theorem establishes between the numbers less than 35 and the pairs (m, n) of numbers in which $m < 5$ and $n < 7$.

3. Counting arguments like the one in the proof of the theorem of this chapter are often imagined in terms of sorting letters (messages in envelopes, not items of the alphabet) into pigeonholes. If the "letters" are the numbers 0–34 and the "pigeonholes" are the pairs (m, n) in which m is one of the numbers 0–4 and n is one of the numbers 0–6, how many letters are there, how many pigeonholes, and how does this image show that the knowledge that each problem $x \equiv m \bmod 5$, $x \equiv n \bmod 7$ has a solution implies that that solution is *unique* for each pair (m, n)?

4. Show that the word "nonzero" can be dropped from the statement of the Chinese remainder theorem (but that if any one a_i is zero, the conclusion becomes trivial).

5. If the requirement that a and b be relatively prime is dropped from the theorem at the beginning of this chapter, the conclusion becomes: (a) *the simultaneous congruences* $x \equiv m \bmod a$ *and* $x \equiv n \bmod b$ *have a solution if and only if* $m \equiv n \bmod d$ where d is the greatest common divisor of a and b, and (b) *if* x *is one solution, then* x' *is a solution if and only if* $x' \equiv x \bmod (ab/d)$. Prove (a) and (b).

6. The simultaneous congruences $x \equiv m \bmod a$ and $x \equiv n \bmod b$ can also be solved by setting $x = qa + m$ and solving the congruence

$qa + m \equiv n \bmod b$ for q. Apply this method to the solution of the simultaneous congruences $x \equiv 10 \bmod 14$ and $x \equiv 20 \bmod 23$.

Computations.

7. Pick two unrelated numbers a and b with 8 digits, and solve $x \equiv 10 \bmod a$ and $x \equiv 20 \bmod b$, if possible, by the method of the text and by the method of Exercise 6.

8. Pick large numbers a, b, and c and find all solutions (x, y) of $ax + b = cy$.

Chapter 7

The Fundamental Theorem of Arithmetic

Because the Euclidean algorithm is practical even for enormous numbers, determining whether *two* given numbers are *relatively prime* is an easy problem, no matter how large the numbers may be. By contrast, determining whether *one* given number is *prime* is much harder when the number is very large (see Chapter 11). Harder still is the problem that gives prime numbers their meaning, the problem of factoring a large number into its prime factors (see Chapter 12). In other words, the factorization that the theorem of this chapter proves exists may be extremely hard to find in practice.

(Before the advent of modern computers, factoring a number with only six or seven digits could be a real challenge. Today's software packages factor numbers with fifteen or twenty digits handily, but numbers with hundreds of digits can still be nearly impossible to factor.)

A number is **composite** if it can be written as a product of two numbers, both of which are greater than 1. A number is **prime** if it is greater than 1 and is not composite. Thus, the numbers 0 and 1 are neither prime nor composite.[1]

[1] In the past there has been some difference of opinion, but today mathematicians agree that 1 should *not* be regarded as prime.

Proposition. *If a is prime and if $bc \equiv 0 \bmod a$, then either $b \equiv 0 \bmod a$ or $c \equiv 0 \bmod a$.*

Proof. Suppose $bc \equiv 0 \bmod a$ but neither $b \equiv 0 \bmod a$ nor $c \equiv 0 \bmod a$, and let d be the greatest common divisor of a and b. The condition $b \not\equiv 0 \bmod a$ implies that $d \neq a$. If d were 1, the augmented Euclidean algorithm would give an equation $1+ua = vb$ and multiplication by c would give $c+uac = vbc$, from which $c+uc\cdot 0 \equiv v\cdot 0 \bmod a$ would follow, contrary to assumption. Therefore, the equation $a = d \cdot \frac{a}{d}$ shows a is not prime. □

The Fundamental Theorem of Arithmetic. *Every number greater than 1 can be written as a product of prime numbers. If two products of prime numbers are equal, say*

$$p_1 p_2 \cdots p_m = q_1 q_2 \cdots q_n$$

where $p_1, p_2, \ldots, p_m$ and $q_1, q_2, \ldots, q_n$ are all prime numbers, then $m = n$ and the two lists $p_1, p_2, \ldots, p_m$ and $q_1, q_2, \ldots, q_n$ of primes are the same, except that they may be ordered differently.

Proof. Let a number $a > 1$ be given. If a is not prime, it is composite and so can be written $a = bc$ where b and c are both greater than 1. If either b or c is not prime, it can be written as a product of two numbers both greater than 1, and the process can be continued, writing a as a product of more and more factors greater than 1, as long as any one of the factors is not prime. This process must terminate with a representation of a as a product of primes before it has been repeated a times, because a product of a factors each of which is greater than 1 must be[2] greater than a.

Two representations $p_1 p_2 \cdots p_m = q_1 q_2 \cdots q_n$ of the same number as a product of primes must be the same, except for the order of the factors, as can be seen in the following way. Since the greatest common divisor of p_1 and $q_1 q_2 \cdots q_n$ is $p_1 > 1$, the lemma of Chapter 6 implies that the greatest common divisor of p_1 and q_i must be greater

[2]This statement is clearly true for $a = 1$. If it is true for some $a > 0$, then it is true for $a + 1$ because a product of $a + 1$ such factors is a number that is at least 2 times a number greater than a, so it is greater than $2a = a + a \geq a + 1$.

than 1 for at least one of the prime factors q_i on the right. By the definition of a prime number, the only divisor of p_1 greater than 1 is p_1, and the same is true of q_i, so the greatest common divisor of p_1 and q_i can be greater than 1 only if $p_1 = q_i$. In short, p_1 must occur among the factors q_i on the right. Rearrange these factors, if necessary, to make $p_1 = q_1$. The original equation then becomes $p_1p_2 \cdots p_m = p_1q_2 \cdots q_n$. Since $m > 1$ if and only if this number is greater than p_1, $m > 1$ if and only if $n > 1$. When this is the case, the same argument can be applied to $p_2p_3 \cdots p_m = q_2q_3 \cdots q_n$ to show that the q's can be rearranged to make $q_2 = p_2$ and that $m > 2$ if and only if $n > 2$. Repetition of this argument m times completes the proof of the theorem. □

Exercises for Chapter 7

Study Questions.

1. Find newspaper articles about the factorization of very large numbers and how certain large numbers have been factored by parceling out the problem in a sophisticated way to many computers collaborating over the worldwide web.

2. Prove using the fundamental theorem of arithmetic that no number A that is not a square can have a rational square root. (In particular, restate this proposition in a way that does not use "rational numbers.")

Computations.

3. Modern computers are so fast that one can actually factor numbers of 10 digits reasonably quickly by brute trial divisions. Implement the following algorithm for doing this and try it out on some large numbers:

```
Input: A number n
Algorithm:
    t = 0
    m = 1
    While m^2 ≤ n
        m = m + 1
        q = 0
```

```
    While n ≥ (q+1)m
      q = q+1
    End
    If n = qm then t = 1
  End
Output: If t = 0 then n is prime, else m is the least prime factor of n
```

4. Speed up the algorithm in Exercise 3 by starting with $m = 0$ and incrementing it by 2 at each step, except that if that incrementation makes it 4, subtract 1 from the result. Do empirical tests to see how effective this modification is in speeding up the algorithm.

5. Further speed up the algorithm of Exercise 3 by replacing the middle "while" loop by a while loop that, instead of adding 1 to q until it no longer satisfies $n \geq qm$, adds to q the largest power of 2 that preserves $n \geq qm$. Again, test empirically how effective this modification is.

Chapter 8

Exponentiation and Orders

Given two nonzero numbers a and b, a **to the power** b, denoted a^b, is by definition the number obtained by multiplying a by itself b times. In everyday experience, raising numbers to powers is difficult for the superficial reason that the answer is normally a *very large* number, so the mere statement of the answer is cumbersome. But for any given c the congruence $a^b \equiv x \bmod c$ has a solution x no larger than c, and there is a simple and altogether practical algorithm for finding it:

```
Input: Nonzero numbers a, b, c
Algorithm:
     Set x = 1, y = a, z = b
     While z > 0
       If z ≡ 0 mod 2
         z = z/2
         y = y^2
         Reduce y mod c
       Else
         z = z − 1
         x = xy
         Reduce x mod c
     End
Output: x
```

As in the case of the augmented Euclidean algorithm, the working of the algorithm can be seen from a table with one column for each of the numbers x, y, and z and with one row for each step of the algorithm. For example, the computation of 2^{90} mod 91 is[1] shown by

x	y	z
1	2	90
1	4	45
4	4	44
4	16	22
4	74	11
23	74	10
23	16	5
4	16	4
4	74	2
4	16	1
64	16	0

with the conclusion that $2^{90} \equiv 64 \bmod 91$. (As the table indicates, $16^2 \equiv 74 \bmod 91$, $4 \cdot 74 \equiv 23 \bmod 91$, $74^2 \equiv 16 \bmod 91$, and $23 \cdot 16 \equiv 4 \bmod 91$, which are easy computations. The other computations are evident.)

At each step, z is decreased—either 1 is subtracted from it, or it is divided by 2. Therefore, the algorithm must eventually reach $z = 0$ and terminate. Each step leaves the value of xy^z modulo c unchanged because $x(y^2)^{z/2} \equiv xy^z \bmod c$ (in the case of steps that divide z by 2) and $xyy^{z-1} \equiv xy^z \bmod c$ (in the case of steps that reduce z by 1) until[2] $z = 1$. Therefore, at the next-to-last step, xy^1 is congruent mod c to $1 \cdot a^b$ and on the last step x itself becomes congruent mod c to a^b. In other words, the output x is indeed congruent to a^b mod c.

The fact that a^b mod c can be computed is not a mere curiosity but an extremely useful tool. (See Chapter 11, for example.) A big part of its usefulness is connected with the solution of:

[1]Note the violation of the rule stated in Chapter 3 that "mod 91" is meaningful only in conjunction with a $\equiv$ sign. Here "mod 91" is used instead in conjunction with the word "computation." The computation does not find 2^{90} itself but only finds it mod 91—it is a "computation mod 91."

[2]Note that y^z is not always defined when $z = 0$. See Exercise 1.

Problem. *Given numbers a and c, both greater than 1, determine all solutions $b > 0$ of $a^b \equiv 1 \bmod c$.*

Proposition. *This problem has a solution b if and only if a is relatively prime to c. When it has a solution, every solution is a multiple of the smallest solution.*

Proof. If the problem has a solution, say $a^b \equiv 1 \bmod c$, then $a^b + sc = 1 + tc$ for some numbers s and t. Since a^b is a multiple of a, the equation $a^b + sc = 1 + tc$ shows that $0 \equiv 1 \bmod [a, c]$ which means that a and c are relatively prime.

For the proof of the converse, assume a and c are relatively prime and consider the first $c + 1$ powers $a, a^2, a^3, \ldots, a^{c+1}$ of a. Each of these $c + 1$ numbers is congruent mod c to one of the c numbers less than c. Therefore, at least two of these powers a^i must be congruent mod c to the same number less than c and therefore must be congruent to each other mod c. In this way one can find nonzero numbers k and l such that $a^k \equiv a^l \bmod c$ and $k < l$. Since a and c are relatively prime, the augmented Euclidean algorithm gives a solution of $1 + uc = va$. Multiplication of $a^k \equiv a^l \bmod c$ by v^k gives $(av)^k \equiv a^l v^k \bmod c$, which is to say $(av)^k = a^{l-k}(av)^k \bmod c$, or simply $1 \equiv a^{l-k} \bmod c$, so the problem has the solution $b = l - k > 0$.

Finally, if b_1 and b_2 are solutions, then their greatest common divisor, call it b_3, is also a solution, because $b_3 + ub_1 = vb_2$ for some numbers u and v, which implies $a^{b_3} \equiv a^{b_3} \cdot 1^u \equiv a^{b_3} \cdot (a^{b_1})^u \equiv (a^{b_2})^v \equiv 1 \bmod c$. Thus, if b_1 is the smaller of the two solutions b_1 and b_2, and if b_2 is not a multiple of b_1, there is a third solution smaller than b_1. In other words, given a solution, trying all smaller numbers either produces a smaller solution or proves that the *only* solutions are the multiples of the given one. □

Definition. *When a and c are relatively prime nonzero numbers, the* **order** *of a mod c is the smallest solution b of $a^b \equiv 1 \bmod c$.*

The proposition states that the solutions $b > 0$ of $a^b \equiv 1 \bmod c$ are the nonzero multiples of the order of a mod c.

Problem. *Given relatively prime nonzero numbers a and c, find the order of a mod c.*

The proof of the proposition shows that the order of a mod c is at most c, so the order can be found by computing a^b mod c for all numbers $b \leq c$. The order of a mod c is simply the smallest b for which the answer is 1. If c is very large, however, this approach to the problem involves far too much calculation to be practical.

The solution of this problem can be *extremely* difficult in some cases, but in other cases it is easy because of a few simple rules that are obeyed by the orders of numbers for a given c. These rules are the subject of the next two chapters, but before you go on to these chapters, you should do lots of examples of small values of c and see if you can discover the rules for yourself.

Exercises for Chapter 8

Study Questions.

1. The definition of a^b requires that a and b both be greater than 0. How would you define it in the case $a > 0$ and $b = 0$? In the case $a = 0$ and $b > 0$? How would you define 0^0? (Note that the explanation of the algorithm in the text would be made a little simpler if it used the definition of a^b in the case $b = 0$, $a \neq 0$.)

Computations.

2. Find the orders mod 13 of all numbers $a < 13$ that are relatively prime to 13.

3. Answer the preceding question for a few numbers other than 13.

4. Find 2^{10} mod 11, 2^{20} mod 21, 2^{30} mod 31, 2^{40} mod 41, 2^{50} mod 51, 2^{60} mod 61, 2^{70} mod 71, 2^{80} mod 81, 2^{90} mod 91, and 2^{100} mod 101 using the above algorithm. You may notice some regularities and near-regularities.

5. Using a programmable calculator and the algorithm of the text, find a^b mod c for some 3-digit numbers a, b, and c.

6. Some of the exercises in the following chapters will require computing a^b mod c for some very large numbers a, b, and c. (Many software systems, including UBASIC, have built-in capabilities for doing such computations. However, programming the algorithm for yourself will give you a better understanding of it.) Compute some examples in which a, b, and c have 10 digits.

Chapter 9

Euler's ϕ-Function

For a positive number c, the number of positive numbers less than or equal[1] to c that are relatively prime to c is called **phi of** c, written[2] $\phi(c)$. It is often called Euler's ϕ-function in honor of the great 18th-century mathematician Leonhard Euler. It plays a central role in determining the orders of numbers mod c.

The values of $\phi(c)$ for small values of c are easily found. For example,

c	$\phi(c)$	c	$\phi(c)$	c	$\phi(c)$	c	$\phi(c)$	c	$\phi(c)$
1	1	11	10	21	12	31	30	41	40
2	1	12	4	22	10	32	16	42	12
3	2	13	12	23	22	33	20	43	42
4	2	14	6	24	8	34	16	44	20
5	4	15	8	25	20	35	24	45	24
6	2	16	8	26	12	36	12	46	22
7	6	17	16	27	18	37	36	47	46
8	4	18	6	28	12	38	18	48	16
9	6	19	18	29	28	39	24	49	42
10	4	20	8	30	8	40	16	50	20

[1] $\phi(c)$ could also be described as the number of positive numbers *less than* c and relatively prime to c, except that this definition would make $\phi(1) = 0$, and it is universally agreed that $\phi(1)$ should be 1.

[2] ϕ is the Greek letter phi.

Study these values of $\phi(c)$ to find the patterns they follow and extend the table up to $c = 100$ using the insights gained.

For example, as is clear from the definition, $\phi(c) = c-1$ when c is prime. Thus, the value of $\phi(c)$ can easily be filled in for all the prime numbers 53, 59, 61, ... between 50 and 100. Also, a comparison of the values of $\phi(c)$ and $\phi(2c)$ gives a simple (but perhaps surprising) relation between the two; the answer depends on whether c is even or odd. Similarly, there is a simple relation between $\phi(c)$ and $\phi(3c)$ that depends on whether c is divisible by 3. A similar relation holds between $\phi(c)$ and $\phi(pc)$ for any *prime* number p. Once this is known, the value of $\phi(c)$ for any c can easily be found for any number c whose factorization into primes is known.

Exercises for Chapter 9

Study Questions.

1. The following rule was hinted at in the chapter. Prove that it is correct:

Proposition. *If p is prime and c is any number, then $\phi(pc) = p\phi(c)$ if p divides c and otherwise $\phi(pc) = (p-1)\phi(c)$.*

[Write numbers less than pc in the form $qc + r$ where $q < p$ and $r < c$ and determine which of them are relatively prime to pc.]

2. From the proposition in the previous question, deduce: *For any number c, $\phi(c) = c(1 - \frac{1}{p_1})(1 - \frac{1}{p_2}) \cdots (1 - \frac{1}{p_k})$ where p_1, p_2, ..., p_k are the* distinct *prime factors of c.* Show how this formula applies in the case of several entries in the table of values of $\phi(c)$. Explain why this formula, despite its appearance, does not really involve fractions.

3. Find $\phi(60)$ using the formula of question 2 and list explicitly the numbers less than or equal to 60 that are relatively prime to 60. (By the way, which, if any, of these $\phi(60)$ numbers are composite?)

4. Prove that if m and n are relatively prime, then $\phi(mn) = \phi(m)\phi(n)$.

Chapter 10

Finding the Order of a mod c

The key to the solution of the problem of determining the order of a mod c when a and c are given relatively prime numbers is the observation that *multiplication by a permutes the numbers counted by $\phi(c)$ in a very special way.*

For example, mod 8, multiplication by 3 permutes the 4 numbers 1, 3, 5, 7 counted by $\phi(8)$ in the following way: $1 \mapsto 3$, $3 \mapsto 1$ $(\equiv 3 \cdot 3 \bmod 8)$, $5 \mapsto 7$ $(\equiv 3 \cdot 5 \bmod 8)$, $7 \mapsto 5$ $(\equiv 3 \cdot 7 \bmod 8)$. The shorthand way to describe this permutation of 1, 3, 5, 7 is as (13)(57). The same shorthand describes the permutation of 1, 3, 5, 7 effected by multiplication by 5 mod 8 as (15)(37) and the one effected by multiplication by 7 mod 8 as (17)(35).

Multiplication by 2 mod 5 effects the permutation $1 \mapsto 2 \mapsto 4 \mapsto 3 \mapsto 1$ of the four numbers counted by $\phi(5)$, for which the shorthand is (1243). Similarly, the permutation of the 6 numbers counted by $\phi(7)$ effected by multiplication by 4 mod 7 is abbreviated (142)(356), the permutation of the 6 numbers counted by $\phi(9)$ effected by multiplication by 2 mod 9 is abbreviated by (124875), and the permutation of them effected by multiplication by 7 mod 9 is abbreviated by (174)(258). (In each case, the number to the right of a number is its image under the permutation, except when there is no number—but a right parenthesis—to the right of the number, in which case the image under the permutation is the number to the right of the corresponding left parenthesis.)

This notation for permutations runs into trouble when the things being permuted include numbers with two digits. For example, multiplication by 2 mod 15 carries $1 \mapsto 2 \mapsto 4 \mapsto 8 \mapsto 1 \equiv 2 \cdot 8 \bmod 15$ and carries $7 \mapsto 14 \mapsto 13 \equiv 2 \cdot 14 \bmod 15 \mapsto 11 \equiv 2 \cdot 13 \bmod 15$. A shorthand way to describe this permutation is to let A, B, and C stand for the 2-digit numbers 11, 13, 14, respectively, that are relatively prime to 15 so that the permutation can be written $(1248)(7CBA)$. Mod 13, the objects being permuted can be written 1, 2, 3, 4, 5, 6, 7, 8, 9, $D = 10$, $E = 11$, $F = 12$. Multiplication by 2 mod 13 then gives the permutation $(124836FE95D7)$, multiplication by 5 mod 13 gives the permutation $(15F8)(2DE3)(4796)$, and so forth.

Theorem. *The permutation of the numbers counted by $\phi(c)$ that is effected by multiplication by $a \bmod c$, where a is any one of them, partitions them into cycles of equal length.*

(Multiplication by 1 mod c is the identity, which can be regarded as the permutation that partitions the numbers counted by $\phi(c)$ into $\phi(c)$ cycles of length 1.)

In other words, in the shorthand way of writing permutations that is described above, the $\phi(c)$ numbers being permuted are written as sets of equal size between parentheses, such as $(174)(258)$ or (124875) or $(15F8)(2DE3)(4796)$. In particular, $\phi(c)$ is a product of two factors—the number of cycles times the number of items in each cycle.

Proof. Let $c > 0$ be given, and let a be a given number relatively prime to c. For any b relatively prime to c, the **orbit of b under multiplication by** $a \bmod c$ is the set of numbers less than c and relatively prime to c that are congruent to $b \cdot a^i \bmod c$ for some i. In terms of the above shorthand for permutations, the orbit of b under multiplication by $a \bmod c$ is simply the set of numbers included in the parentheses that include b. The theorem states that these orbits all have the same size.

The theorem will be proved by showing that for any b *the length of the orbit of b is the order of a* mod c, so the length of the orbit of b does not depend on b.

The orbit of b under multiplication by a mod c clearly contains *at most* r distinct numbers, where r is the order of a mod c, because $b \equiv ba^r \equiv ba^{2r} \equiv ba^{3r} \equiv \cdots \bmod c$ and, more generally, any two numbers in the list b, ba, ba^2, ba^3, $\ldots$ that are r steps apart are congruent mod c.

What is to be shown, then, is that $ba^i \not\equiv ba^{i+j} \bmod c$ whenever $0 \le i < j < r$. If this statement were false, there would be a number k in the range $0 < k < r$ for which $ba^i \equiv ba^{i+k} \bmod c$. But $ba^i \equiv ba^{i+k}$ is impossible for $0 < k < r$ because a and b are both relatively prime to c, so one would be able to multiply this congruence once by the reciprocal of b mod c and i times by the reciprocal of a mod c to find $1 \equiv a^k \bmod c$, which would be contrary to the definition of r as the smallest positive solution of $a^r \equiv 1 \bmod c$. □

Corollary. *If a is relatively prime to c, then $a^{\phi(c)} \equiv 1 \bmod c$. Otherwise stated, the order of a mod c divides $\phi(c)$.*

Deduction. Say that multiplication by a mod c is a permutation that consists of e cycles, each of length f. Since f repetitions of a cyclic permutation of length f returns each item to its original place, f repetitions of multiplication by a mod c is the identity. In other words, $a^f \equiv 1 \bmod c$. Therefore, $a^{\phi(c)} \equiv a^{ef} \equiv (a^f)^e \equiv 1^e \equiv 1 \bmod c$, as was to be shown. □

Problem. *Given relatively prime nonzero numbers a and c, find the order of a mod c.*

In practice, the solution of this problem can be difficult, but it is easy *if $\phi(c)$ can be factored into primes* because then the factors of $\phi(c)$ can be enumerated and the order of a mod c can be determined by a few simple tests.

Specifically, it is known that $a^{\phi(c)} \equiv 1 \bmod c$ and the question is whether any factor of $\phi(c)$ smaller than $\phi(c)$ has this property. If so, then some number of the form $\phi(c)/p$, where p is a prime factor of $\phi(c)$, must have the property, because every factor of $\phi(c)$ less than $\phi(c)$ divides a number of this form—it leaves out at least one prime factor of $\phi(c)$. Thus, one tests whether $a^{\phi(c)/p} \equiv 1 \bmod c$ for each prime factor p of $\phi(c)$. If this congruence is not satisfied for any prime factor p of $\phi(c)$, then $\phi(c)$ is the order of $a \bmod c$. But if a prime factor p of $\phi(c)$ is found for which $a^{\phi(c)/p} \equiv 1 \bmod c$, one can then begin trying exponents which are $\phi(c)/p$ divided by one of *its* prime factors, and so forth. Eventually (and usually rather soon) one will find a factor f of $\phi(c)$ for which $a^f \equiv 1 \bmod c$ but $a^{f/p} \not\equiv 1 \bmod c$ for each prime factor p of f. This f is the order of $a \bmod c$.

Traditionally, the fundamental fact of elementary number theory expressed by the corollary has the unsatisfactory name "Euler's generalization of Fermat's theorem." The case in which c is prime is "Fermat's theorem" itself. That is, Fermat's theorem is the statement that $a^{p-1} \equiv 1 \bmod p$ when p is prime and $a \not\equiv 0 \bmod p$. Or, better, "Fermat's theorem" is the congruence obtained by multiplying $a^{p-1} \equiv 1 \bmod p$ by a so that the statement is also true when $a \equiv 0 \bmod p$:

Fermat's Theorem[1]. *If p is prime, then $a^p \equiv a \bmod p$ for all numbers a.*

Exercises for Chapter 10

Study Questions.

1. Reconsider the computations in Exercise 4 of Chapter 8. Which of the numbers 11, 21, 31, 41, 51, 61, 71, 81, 91 are prime?

2. Fermat's theorem can be regarded as a statement about *binomial coefficients.* Construct Pascal's triangle showing the coefficients

[1] It has become fashionable in recent years to call this theorem "Fermat's Little Theorem." It is a crucial fact of elementary number theory and should not be belittled.

in the expansion of $(a+b)^n$ for $n = 2, 3, \ldots, 8$. For these values of n, which of them are divisible by n? (a) Formulate what you find in the case of *prime* exponents as a congruence involving $(a+b)^p$ when p is prime. (b) Prove this congruence. [Hint: The formula for binomial coefficients as fractions in which numerator and denominator involve factorials is useful.] (c) Deduce Fermat's theorem from it.

3. The formula $x^k - 1 = (x-1)(x^{k-1} + x^{k-2} + \cdots + x + 1)$ proves that $2^n - 1$ is composite whenever n is composite. (For example, 2^{15} has the factorization $(2^5 - 1)(2^{10} + 2^5 + 1)$.) If n is prime, $2^n - 1$ may be, and often is, prime. Such primes are called *Mersenne primes.* They are a popular topic because they are so easily proved to be prime (when they *are* prime). Use the corollary of this chapter to prove that *if n is prime and if p is a prime factor of $2^n - 1$, then $p \equiv 1 \bmod n$.* Moreover, to determine whether p divides $2^n - 1$, one only needs to do the simple computation of $2^n \bmod p$. Clearly $2^2 - 1 = 3$, $2^3 - 1 = 7$, $2^5 - 1 = 31$ are all prime. That $2^7 - 1 = 127$ is prime follows from the above ideas *without computation.* Use these ideas to determine whether the next few numbers $2^n - 1$ for $n = 11, 13, 17, \ldots$ are prime.

Computations.

4. (a) Find the permutation of the numbers less than 25 and relatively prime to 25 that is effected by multiplication by 4 mod 25. (Use the letters a, b, c, d for 11, 12, 13, 14, the letters e, f, g, h for 16, 17, 18, 19, and the letters i, j, k, l for 21, 22, 23, 24.) (b) Find the permutation that is effected by multiplication by 2 mod 25. (c) Show that the answer to (a) is the square of the answer to (b). (d) What is the square of the answer to (a)? (e) What is the square of the answer to (d)?

5. For large relatively prime numbers a and c, find the order of $a \bmod c$. For a start, you might try finding the orders of 99, 100, and 101 mod 221, but you can also answer the question for numbers with several digits, using a computer to do exponentiations and factorizations. (For factorizations, use the program in the exercises of Chapter 7.)

Chapter 11

Primality Testing

Fermat's theorem has the surprising corollary that a number can be proved to be composite without any investigation of its factors. For example, as was shown in Chapter 8, $2^{90} \equiv 64 \bmod 91$. The theorem of the last chapter proves that if 91 were prime, 2^{90} would be 1 mod 91, so 91 must be composite. Of course, 91 is more easily proved to be composite by giving the explicit factor 7, but for very large odd numbers n it is usually easier to compute $2^{n-1} \bmod n$ than it is to look for factors.

For example, many trial divisions are necessary to find the prime factorization of 1022117, but the relatively simple computation of $2^{1022116} \equiv 467183 \bmod 1022117$ is all that is needed to determine that 1022117 is composite.

In the same way, computation of $3^{1022116} \equiv 537878 \bmod 1022117$ proves that 1022117 is composite, as does computation of $a^{1022116} \bmod 1022117$ for any number a for which $a^{1022116} \not\equiv 1 \bmod 1022117$. A single a for which $a^{n-1} \not\equiv 1 \bmod n$ suffices to prove that n is composite.

Primality Test. *Given a number n, choose a number a in the range $1 < a < n$ and compute $a^{n-1} \bmod n$. If the result is not $1 \bmod n$, then n is composite.*

This test is *incomplete* in the sense that it may not determine whether n is prime. It *tests* the primality of n in the sense that hardship tests a person's character. Hardship may prove that character is lacking, and the above test may prove that a number is not prime, but a person of poor character may withstand hardship and a number may pass the test without being prime. If one only needs to make a highly reliable guess as to whether a given number is prime, this test is very useful because experience shows that most composite numbers are *proved* to be composite by just one test of this type; a number that is not proved composite by a half a dozen or so such tests is very likely not composite—which is to say that it is probably prime.

Again, the test can prove that a number is composite, but it can never prove that a number is prime.

Another way to describe the situation is to say that $a^{n-1} \equiv 1 \bmod n$ is a *necessary* condition for n to be prime. It is a *strong* necessary condition in the sense that relatively few composite numbers go undetected by it, but it is not a sufficient condition; many examples do exist of pairs of numbers a and n in which n is composite, $1 < a < n$, and $a^{n-1} \equiv 1 \bmod n$.

The test is greatly strengthened by the following simple observation:

Lemma. *If p is prime and if x is a number satisfying $x^2 \equiv 1 \bmod p$, then either $x \equiv 1 \bmod p$ or $x \equiv -1 \bmod p$.*

Proof. Because $x^2 \equiv 1 \bmod p$, $x \not\equiv 0$. Therefore, $x - 1$ is a number and the product $(x-1)(x+1) = x^2 - 1$ is divisible by p. Since p is prime, p divides at least one of the factors on the left, as was to be shown. □

Since there is never any doubt about whether an even number is prime, the above test can be strengthened to:

Primality Test. *Given an odd number n, choose a number a in the range $1 < a < n$ and compute $a^{(n-1)/2} \bmod n$. If the result is not $\pm 1 \bmod n$, then n is composite.*

This revised test is not only simpler in the sense that it calls for raising a to a lower power, it is also stronger in the sense that it catches some composite numbers that elude the first test. For example, $3^{3300} \equiv 4509 \bmod 6601$ so 6601 is composite, even though $3^{6600} \equiv 1 \bmod 6601$.

The lemma leads to an even stronger test in many cases, because when $a^{(n-1)/2} \equiv 1 \bmod n$ *and* $\frac{n-1}{2}$ is even, one knows that $a^{(n-1)/4}$ must be $\pm 1 \bmod n$ if n is prime. Moreover, if $a^{(n-1)/4}$ is found to be $1 \bmod n$ and if $\frac{n-1}{4}$ is even, then $a^{(n-1)/8}$ must be $\pm 1 \bmod n$ if n is prime, and so forth.

When this method is applied to $n = 6601$ with $a = 2$, one finds first $2^{3300} \equiv 1 \bmod 6601$, which leads one to compute $2^{1650} \bmod 6601$. Since in fact $2^{1650} \equiv 4509 \bmod 6601$, 6601 is composite (as was found above using $a = 3$ instead of $a = 2$).

The primality test that follows from these observations, which is called **Miller's test**, is described by the following algorithm:

Miller's Test for Base a.

```
Input: n, a      (1 < a < n and n is odd)
Algorithm:
    t = 1
    e = (n − 1)/2
    While t = 1
      Compute u ≡ a^e mod n
      If u ≡ −1 mod n then t = 0
      Elseif u ≡ 1 mod n and e is even then e = e/2
      Elseif u ≡ 1 mod n then t = 0
      Else t = 2
    End
Output: If t = 2 print "n is composite" else print "n passes Miller's test for
    the base a"
```

For n to pass Miller's test (for t to become 0), n must satisfy one of two conditions:

(1) the congruence $a^{(n-1)/2^i} \equiv -1 \bmod n$ holds for some $i > 0$ for which 2^i divides $n - 1$, or

(2) $a^{(n-1)/2^i} \equiv 1 \bmod n$ when i is the number of times 2 divides $n - 1$.

Otherwise, there is an even exponent m for which $a^m \equiv 1 \bmod n$ and $a^{m/2} \not\equiv \pm 1 \bmod n$, so n must be composite.

(Note that Miller's test never asks whether $a^{n-1} \equiv 1 \bmod n$. If this condition fails, then (1) or (2) must fail.)

Fact: The number $3215031751 = 151 \cdot 751 \cdot 28351$ is the only composite number less than 2.5×10^{10} that passes Miller's test for all of the bases 2, 3, 5, and 7. Miller's test for base 11 proves it is composite.

Considering the number of composite numbers less than 2.5×10^{10}, this fact shows how strong Miller's test is, even though a composite number can occasionally slip by it undetected.

Exercises for Chapter 11

Study Questions.

1. Does it make sense to take the base a in Miller's test to be a prime?

2. If $(n-1)/2$ is even and $a^{(n-1)/2} \not\equiv \pm 1 \bmod n$, Miller's test requires computing $a^{(n-1)/4} \bmod n$. If it had been known at the outset that $a^{(n-1)/4}$ would be needed, it would have been more efficient to compute it first and then to find $a^{(n-1)/2} \bmod n$ by squaring. For this reason, Miller's test is often performed by first computing $a^{(n-1)/2^i} \bmod n$ where i is the number of times that 2 divides $n-1$ and repeatedly squaring to find $a^{(n-1)/2^{i-1}} \bmod n$, $a^{(n-1)/2^{i-2}} \bmod n$, ..., $a^{(n-1)/2} \bmod n$. Express this test in the form of an algorithm like the one in the text.

Computations.

3. The number 12801 is composite because it is divisible by 3. Miller's test for the base 2 and Miller's test for the base 5 both prove it is composite, but both require rather a lot of computation. Do the computations.

4. Choose a 3-digit number xxx, and for each odd number n between xxx00 and xxx50, use an implementation of the exponentiation algorithm to compute $2^{(n-1)/2} \bmod n$. If the answer is not $\pm 1 \bmod n$

(remember that -1 does not appear as -1 but as $n - 1$ in the output of the exponentiation), then n is composite. In all likelihood, a number of values of n will remain. For each of them, complete Miller's test for the base 2. (It is already complete if $\frac{n-1}{2}$ is odd or if $2^{(n-1)/2} \equiv -1 \bmod n$.) Some values of n will probably remain. For each of them, try to prove they are composite by using Miller's test for the base 3 and the base 5. (It is unlikely that this will succeed, because composite numbers that pass Miller's test for the base 2 are—especially in this range—rare.) Finally, use a factorization program (feasible in this range) to verify that the remaining numbers are all prime (or—very unlikely—to identify a composite number that passes Miller's test for the bases 2, 3, and 5).

5. Try to find, by trial-and-error, a composite number n that passes Miller's test for some base a. (Not easy. In fact, there is little hope of success without using a computer to try hundreds of cases. As the case of 3215031751 suggests, a good way to proceed is to look for a number n that is proved composite by Miller's test for one base a but passes the test for another base.)

Chapter 12

The RSA Cipher System

In this information age, everyone understands that all forms of information, from recorded sounds to pictures of distant galaxies, can be expressed as sequences of zeros and ones—that is, as *numbers*. It is natural, therefore, to restate the problem of encoding and decoding messages—which we normally think of as sequences of letters—as the problem of encoding and decoding numbers.

Specifically, let the problem be imagined in the following way: We are at one end of a phone line, and we are eager to know a number that is in the possession of the person, call him Deep Throat, at the other end of the line, but we are convinced that numbers sent over the line can be monitored by other parties who are as eager to know the number as we are to keep it secret from them. For the sake of definiteness, let us say that the secret number has 99 digits.

The RSA system, invented in 1978 by R. L. Rivest, A. Shamir, and L. M. Adleman, gives us the following way to accomplish our goal:

We first find two 51-digit prime numbers; call them p and q. This is a substantial computation, but a relatively easy one using Miller's test and a computer program capable of adding and multiplying very large integers. We will just pick a 51-digit number that is not divisible by 2, 3, or 5 and apply Miller's test for the bases 2, 3, and 5. If the tests show it is composite, we will add 30 and try again. Soon, experience shows, we will find a number that passes these three tests. Chances are, it is prime, but, for safety's sake, we will apply Miller's test for several other bases a, before deciding to choose this number as p. We will follow the same procedure to choose q, another 51-digit prime that will resemble p in no other way.

Next, we will compute the product of p and q; call it n. This n will have 100 or 101 digits. (Since p and q are between 10^{50} and 10^{51}, their product is between 10^{100} and 10^{102}.) We will keep p and q as our most carefully guarded secret, but we will send n to Deep Throat over the phone line that connects us. We will also send him another 100 digit number e chosen at random subject to the sole condition that it be relatively prime to $\phi(n)$.

(We know $\phi(n)$ because

$$\phi(n) = \phi(pq) = \phi(p)\phi(q) = (p-1)(q-1) = n - p - q + 1.$$

Thus, we can quickly determine whether any given e is relatively prime to $\phi(n)$ using the Euclidean algorithm.)

Our instruction now to Deep Throat is: Take the secret number, call it P (it is the "plain text"), raise it to the power e mod n, say $C \equiv P^e \bmod n$ (C is the "cipher text" or "code text"), and send us C over the phone line.

We don't care that our adversaries can intercept the number C—just as they could have intercepted n and e—because in their ignorance of p and q they can't do what we will do to find the secret number P, which is to use the augmented Euclidean algorithm to find a reciprocal of e mod $\phi(n)$, say $re \equiv 1 \bmod \phi(n)$, and to raise C to the power r mod n. The resulting number—assuming, of course, that it is reduced mod n—will be the secret number!

That is:

Proposition. *If p and q are distinct primes, if $n = pq$, and if $re \equiv 1 \bmod \phi(n)$, then $(P^e)^r \equiv P \bmod n$.*

Proof. To prove $P^{er} \equiv P \bmod n$, it will suffice (by virtue of the Chinese remainder theorem) to prove that $P^{er} \equiv P \bmod p$ and $P^{er} \equiv P \bmod q$. Since the conditions are symmetric in p and q, it will suffice to prove just $P^{er} \equiv P \bmod p$.

The congruence $P^{er} \equiv P \bmod p$ is obviously true when $P \equiv 0 \bmod p$. When $P \not\equiv 0 \bmod p$, Fermat's theorem implies $P^{p-1} \equiv 1 \bmod p$. When s is defined by $re \equiv s\phi(n) + 1$, one finds then that $P^{re} \equiv P^{s\phi(n)+1} \equiv P^{s(p-1)(q-1)} \cdot P \equiv (P^{p-1})^{s(q-1)} \cdot P \equiv 1 \cdot P \equiv P \bmod p$, which completes the proof. □

Note that a slight modification of the proof shows that $re \equiv 1 \bmod \phi(n)$ implies $P^{re} \equiv P \bmod n$ *whenever n is a product of distinct primes,* even when there are more than two of them. If we are mistaken and the $n = pq$ we have chosen is not the product of two primes p and q, it will still very likely be true that $re \equiv 1 \bmod \phi(n)$ implies $P^{re} \equiv P \bmod n$ (it is very unlikely that p or q would have a repeated prime factor), *but* it will not be true that $\phi(n) = (p-1)(q-1)$, so the reciprocal r of $e \bmod (p-1)(q-1)$ that is easy for us to compute will not in fact be the reciprocal of $e \bmod \phi(n)$ that is needed to find P. Therefore, the system will not work if either p or q fails to be prime. If we had the bad luck to have chosen, despite our precautions, a value that was not prime, the message we received would be garbled, and we would need to try again with a new p and a new q. If the situation is so critical that there can be no second chance, we would need to use more—and more sophisticated—tests of the primality of p and q.

The real question is whether our assumption is correct that our adversaries won't be able to find the magic number r we used to decode the message. We were able to find it using the augmented Euclidean algorithm because we knew the prime factors p and q of n and therefore knew $\phi(n)$. Our adversaries could do the same if they could factor n, but it is generally believed that this factorization problem, though of course it can be solved in principle, is *extremely* difficult and requires an enormous investment on the part of anyone trying to intercept the message. The needed investment can be increased by increasing the sizes of p and q or by adding a second level of encryption—that is, by first transmitting as a secret number the code that will be used to transmit the actual secret number.

(Note that knowing $\phi(n)$ is virtually the same as knowing the factorization $n = p \cdot q$ of n because p and q are the roots of the quadratic polynomial $(x-p)(x-q) = x^2 - (p+q)x + pq = x^2 - (n - \phi(n) + 1)x + n$ and can therefore be found using the quadratic formula—square roots are easy—once n and $\phi(n)$ are known.)

If anyone has secretly devised a method of finding r given n and e without the huge investment that is generally believed to be necessary, the secret is well guarded.

Warning: Cryptography is not pure mathematics; codes are often broken using subtle clues that are unrelated to the mathematical problem the code poses. This book is not the place for an examination of such subtleties.

Exercise for Chapter 12

Let the secret number be $P = 12345678909876554321$. Use Miller's test and trial-and-error to find two (very probably) prime numbers p and q with ten digits. Use trial-and-error to find a 15-digit number e relatively prime to $\phi(pq) = (p-1)(q-1)$. Momentarily take the role of Deep Throat—who is given the values of e and $n = pq$ (but not of p and q)—and compute $C \equiv P^e \bmod n$. Then satisfy yourself that you can find $P \equiv C^r \bmod n$ by using your knowledge of the factorization of n to determine the needed r.

Chapter 13

Primitive Roots mod p

By Fermat's theorem, the order mod p of any a that is relatively prime to p divides $p-1$. In particular, $p-1$ is the largest possible order mod p of any number relatively prime to p. One finds in fact that for any prime p the order of at least one a mod p is $p-1$. For example, the orders mod 7 of 1, 2, 3, 4, 5, and 6 are 1, 3, 6, 3, 6, and 2, respectively, so the maximum order 6 occurs twice.

It seems to have been well known, many years before Gauss gave the first proof, that *for every prime p there are numbers relatively prime to p whose orders mod p have the maximum value $p-1$*. Such numbers are called **primitive roots mod p**. It was even known that *the number of primitive roots mod p is exactly $\phi(p-1)$*. For example, there are $\phi(6) = 2$ primitive roots mod 7, namely, 3 and 5. (In counting the primitive roots, it is natural to count just those that are less than p. Strictly speaking, 10 is a primitive root mod 7, but it is to be regarded in this context as being the same as 3.)

Gauss says: "This theorem gives an excellent example of how much caution must be taken in the theory of numbers not to assume things are true that in fact have not been proved" [**G**, Art. 56]. He goes on to say that one of his predecessors, Lambert, had stated that there are $\phi(p-1)$ primitive roots mod p without even mentioning the need for a proof, and that his most illustrious predecessor, Euler, recognized that a proof was necessary but failed to give an entirely satisfactory one. Gauss himself gave two proofs in his *Disquisitiones Arithmeticae.* His second proof (in Article 54 of the book) is perhaps a little less elegant than his first, but it is more algorithmic—it tells exactly how to find a primitive root.

Theorem. *For any prime p, there is a primitive root mod p.*

Gauss's proof in Article 54 depends on a simple lemma that will be proved in the next chapter:

Lemma. *If p is a prime number and if n is any positive number, there are at most n numbers a less than p for which $a^n \equiv 1$ mod p.*

To put it more colloquially, the number of nth roots of 1 mod p is at most n. More generally, a polynomial of degree n (in the case at hand the polynomial is $x^n - 1$) has at most n distinct roots mod p.

Proof of the theorem. If a primitive root mod p is known, one can find a number whose order mod p is any given factor of $p - 1$. Specifically, if g is a primitive root mod p and f is a factor of $p - 1$, then $g^{(p-1)/f}$ has order f mod p. (Its fth power is $g^{p-1} \equiv 1 \bmod p$ and a lower power of $g^{(p-1)/f}$ is g raised to a power less than $p - 1$ so it cannot be 1 mod p by the defining property of g.) Gauss's proof goes the other way: It constructs elements of order f for certain types of factors of $p - 1$ and uses them to construct a primitive root mod p. Specifically, it constructs elements whose order mod p is q^e for each prime factor q of $p - 1$, where e is the number of times q divides $p - 1$.

First, for each distinct prime factor q of $p - 1$ find a number a in the range $1, 2, \ldots, p-1$ for which $a^{(p-1)/q} \not\equiv 1 \bmod p$. By the lemma, at most $(p - 1)/q$ numbers in this range *fail* to have this property, so such a number is sure to be found after at most $\frac{p-1}{q} + 1$ trials.

Next, for each such q let a number b be determined mod p by the condition $b \equiv a^{(p-1)/q^e} \bmod p$, where a is the number found for this q in the first step and where e is the number of times q divides $p - 1$. (If $e = 1$, as will be the case for most if not all of the prime divisors q of $p - 1$, the number $b \equiv a^{(p-1)/q}$ will already have been found in the course of finding a.)

It will be shown that the *product,* call it g, of the numbers b found in this way, with one factor of g for each prime factor q of $p-1$, is a primitive root mod p.

The factor b of g corresponding to the prime q satisfies $b^{q^e} \equiv 1 \bmod p$, where e is the number of times q divides $p-1$, because $b^{q^e} \equiv a^{p-1} \equiv 1 \bmod p$ for the corresponding a. Therefore, the order of $b \bmod p$ divides q^e, which means that the order of $b \bmod p$ is q^j for some $j \leq e$. If j were not equal to e then the order of b would divide q^{e-1}, which is impossible because b to the power q^{e-1} is a to the power $\frac{p-1}{q^e} \cdot q^{e-1} = \frac{p-1}{q}$ and a was chosen to make $a^{(p-1)/q} \not\equiv 1 \bmod p$. In short, b must have order $q^e \bmod p$.

For a prime factor Q of $p-1$, it follows that $b^{(p-1)/Q} \equiv 1 \bmod p$ if and only if Q is *not* the prime q used to construct b, because the order q^e of $b \bmod p$ fails to divide $\frac{p-1}{Q}$ if and only if division by Q removes one of the e occurrences of q in the factorization of $p-1$. Thus $g^{(p-1)/Q}$ is a product of factors $b^{(p-1)/Q}$, all but one of which are $1 \bmod p$, but the remaining one, which corresponds to the prime factor Q of $p-1$, is *not* $1 \bmod p$.

Therefore $g^{(p-1)/Q} \not\equiv 1 \bmod p$ for each prime factor Q of $p-1$. Since the order of $g \bmod p$ divides $p-1$, it must therefore be $p-1$, because every other divisor of $p-1$ divides $(p-1)/Q$ for at least one prime factor Q of $p-1$. Thus, g is a primitive root mod p, as was to be shown. □

Corollary. *For any prime p, the number of numbers a in the range $0 < a < p$ whose orders $\bmod p$ are r is 0 if r does not divide $p-1$ and is $\phi(r)$ if r does divide $p-1$. In particular, the number of primitive roots mod p is $\phi(p-1)$.*

Deduction. The statement to be proved is clear in the case $r = 1$—there is just one number less than p whose order mod p is 1, namely, the number 1, and $\phi(1) = 1$. Therefore, assume $r > 1$.

That there are no numbers of order r when r does not divide $p-1$ follows from Fermat's theorem.

Suppose now that r does divide $p-1$, and let g be a primitive root mod p. As was noted above, $g^{(p-1)/r}$ has order $r \bmod p$. It will be shown that if a is one number whose order mod p is r, then a number b has order r mod p if and only if it is of the form $b \equiv a^i \bmod p$ where i is less than r and relatively prime to r. Since there are $\phi(r)$ such values of i less than $p-1$ (of course one counts two values of i as being the same if they are congruent mod $(p-1)$ because then they produce the same value of $a^i \bmod p$), the corollary will follow.

If a has order r mod p, then $a, a^2, a^3, \ldots, a^r$ are distinct mod p and all of them satisfy $(a^i)^r \equiv 1 \bmod p$. By the lemma above, which is to be proved in Chapter 14, there are at most r distinct numbers mod p with this property, so $x^r \equiv 1 \bmod p$ implies $x \equiv a^i \bmod p$ for some i. By the same token, if a^i has order r, then, since $a^r \equiv 1 \bmod p$, a must be a power of a^i, say $a = (a^i)^j$. Then $a \equiv a^{ij} \bmod p$, which implies $1 \equiv a^{ij-1} \bmod p$ so the order r of $a \bmod p$ divides $ij - 1$, which is to say $ij \equiv 1 \bmod r$. In particular, i is relatively prime to r. Thus, every element of order r has the form a^i where i is at most r and is relatively prime to r. Conversely, if i is relatively prime to r, then the augmented Euclidean algorithm gives a solution (u, v) of $1 + ur = vi$ so $a \equiv a \cdot (a^r)^u \equiv (a^i)^v \bmod p$, so every power of a is a power of a^i. Since a has r distinct powers, so must a^i, and the order of a^i, which is obviously at most r, cannot be less than r, as was to be shown. □

Exercises for Chapter 13

Study Questions.

1. Show by repeated doubling of 1 that 2 is a primitive root mod 13. Use this fact to find the numbers whose orders mod 13 are r for $r = 12, 6, 4, 3, 2$, and 1.

2. Make use of the corollary to prove the interesting identity

$$\sum_{d|(p-1)} \phi(d) = p - 1$$

where the sum is over all divisors d of $p-1$. (For example, when $p = 13$, the identity is $\phi(1) + \phi(2) + \phi(3) + \phi(4) + \phi(6) + \phi(12) = 12$, which is to say $1 + 1 + 2 + 2 + 2 + 4 = 12$.) Verify the formula for a few other primes p.

3. Given one number whose order mod p is 25, describe a construction that finds *all* numbers whose order mod p is 25. How many are there? In this case, how many elements of order 5 are there and how can they be found?

Computations.

4. Find a primitive root mod 101 and use it to find one number whose order mod 101 is 25 and one number whose order mod 101 is 5.

5. Use Gauss's method to find a primitive root mod 257. How many are there? Is 3 one of them? Is 11? Give a simple test that tells whether a given number is a primitive root mod 257.

6. Use Miller's test to find a 6-digit (very probably) prime p whose last two digits are 01. Factor $p - 1$ and use Gauss's method to find, for each prime factor q of $p - 1$, a number whose order mod p is q^e where e is the number of times q divides $p - 1$. Then find a primitive root mod p by finding their product mod p.

7. For the primes $p = 37$, 41, 43, determine whether there are numbers whose orders mod p are 4 and, if so, find all such numbers.

Chapter 14

Polynomials

The method of algebra, of performing computations that involve *letters* as well as numbers, has a place in number theory too. More properly, the letters are called *indeterminates.* They may eventually be given numerical values—and they may be unknowns of the problem which will eventually be determined at the end of the calculations—but one computes with them without knowing anything about them. The only assumption is that they obey the same computational rules—associativity, distributivity, commutativity—that numbers do.

In this chapter, computations that involve *one* letter will be used to prove the lemma of Chapter 13. The natural choice for that one letter is x. If you start with x and allow addition and multiplication, possibly involving numbers, you will be dealing with *polynomials in x,* which is to say expressions of the form $c_0x^n+c_1x^{n-1}+c_2x^{n-2}+\cdots+c_n$ in which n is a number and the "coefficients" $c_0, c_1, \ldots, c_n$ are also numbers. Polynomials are added and multiplied according to the usual rules—which follow from the associative, distributive, and commutative laws just mentioned—and the result is another polynomial. Note that numbers are included among polynomials—namely, as polynomials in which $n = 0$, the so-called constant polynomials.

The concept of congruence mod m for a number m extends to polynomials in an obvious way. Given a number m, two polynomials are said to be *congruent mod m* if they can be made equal by adding polynomials of the form mx^i ("multiples of m" when the meaning of this phrase is extended to include multipliers that involve x) to one or the other or both of them. Otherwise stated, two polynomials $c_0x^n + c_1x^{n-1} + c_2x^{n-2} + \cdots + c_n$ and $c_0'x^n + c_1'x^{n-1} + \cdot + c_n'$ are congruent mod m if $c_i \equiv c_i'$ mod m for each i.

The theorem below deals with polynomials in which $c_0 = 1$. Such polynomials are called **monic**.

As one easily sees, computations with polynomials mod m make sense in the same way that computations with numbers mod m do: Sums of polynomials are congruent mod m when the summands are congruent mod m, and the same is true of products.

With polynomials, there is a new type of operation, along with addition and multiplication, which is the operation of *evaluation.* If $f(x)$ is a polynomial and a is a number, then the *value* of $f(x)$ when $x = a$, denoted $f(a)$, is simply the number that results when you replace the letter x with the number a and compute all the powers, products, and sums indicated by the polynomial. This operation is "consistent with addition and multiplication" in the sense that $f(a) + g(a)$ is the value of $f(x) + g(x)$ when $x = a$, and $f(a)g(a)$ is the value of $f(x)g(x)$ when $x = a$ because these rules follow from the commutative, associative, and distributive laws of arithmetic.

The main step in the proof of the lemma of Chapter 13 is the following limited sort of *division with remainder* of polynomials.

Theorem. *Let $f(x)$ be a monic polynomial $x^n + c_1x^{n-1} + c_2x^{n-2} + \cdots + c_n$ in which $n > 0$, let m be a modulus greater than 0, and let b be a number. The congruence*

$$(1) \qquad f(x) \equiv (x - b)q(x) + r \quad \text{mod } m$$

determines a monic polynomial $q(x) = x^{n-1} + q_1x^{n-2} + \cdots + q_{n-1}$ mod m and a number r mod m in the sense that there are solutions $q(x)$ and r of this problem, and all solutions are congruent mod m.

(Although, strictly speaking, $x-b$ is not a "polynomial" according to the above definition, the meaning of congruence (1) should be clear, because minus signs always make sense in congruences. Instead of writing $x - b$, one could write $x + a$ where a is a number for which $a + b \equiv 0 \bmod m$.)

Proof. The congruence $f(x) \equiv (x - b)q(x) + r \bmod m$ amounts to n congruences $c_1 \equiv q_1 - b \bmod m$, $c_2 \equiv q_2 - q_1 b \bmod m$, ... , $c_{n-1} \equiv q_{n-1} - q_{n-2}b \bmod m$, and $c_n \equiv -q_{n-1}b + r \bmod m$ when one observes that the coefficients of corresponding powers of x must be congruent mod m. When these congruences are rewritten in the form $q_1 \equiv c_1 + b \bmod m$, $q_2 \equiv c_2 + q_1 b \bmod m$, ... , $q_{n-1} \equiv c_{n-1} + q_{n-2}b \bmod m$, and $r \equiv c_n + q_{n-1}b$, they make it possible to determine $q_1, q_2, \ldots, q_{n-1}$, and r successively mod m, so $f(x)$ and b determine $q(x)$ and r mod m. Conversely, if $q(x)$ and r are found using these congruences, then the original congruences are satisfied, which is to say that $f(x) \equiv (x - b)q(x) + r \bmod m$. □

Corollary. *The lemma of Chapter 13. More generally, for a given prime number p and a given monic polynomial of degree n, at most n numbers b less than p satisfy $f(b) \equiv 0 \bmod p$.*

Deduction. Let the number of solutions $b < p$ of $f(b) \equiv 0 \bmod p$ be called the "number of roots of $f(x)$ mod p." Thus, the problem is to show that a monic polynomial of degree n cannot have more than n roots mod p.

Let b be a root of $f(b) \equiv 0 \bmod p$, where p is prime, and let the theorem be used to determine a polynomial $q(x)$ of degree $n - 1$ and a remainder r for which $f(x) \equiv (x - b)q(x) + r \bmod p$. Then $f(b) \equiv (b - b)q(b) + r \bmod p$ shows that $0 \equiv r \bmod p$ so r can be taken to be zero. Any root b' of $f(x)$ satisfies $(b' - b)q(b') \equiv f(b') \equiv 0 \bmod p$. If $b' \not\equiv b \bmod p$, then the factor $b' - b$ is not congruent to zero mod p, so, since p is prime, the congruence $(b' - b)q(b') \equiv 0 \bmod p$ implies $q(b') \equiv 0 \bmod p$.

In short, each root of $f(x)$ mod p other than b is a root of $q(x)$ mod p, so the existence of a monic polynomial $f(x)$ of degree n with more than n distinct roots mod p would imply the existence of a monic polynomial $q(x)$ of degree $n - 1$ with more than $n - 1$ distinct roots mod p.

Repetition of this argument $n-1$ times shows that the existence of a monic polynomial of degree n with more than n distinct roots mod p would imply the existence of a polynomial $x+c$ that had more than one root mod p, which is absurd because $b+c \equiv 0 \equiv b'+c \bmod p$ implies $b \equiv b' \bmod p$. Therefore, a monic polynomial of degree n can have at most n distinct roots mod p, as was to be shown. □

Exercises for Chapter 14

1. (a) Find the number less than 13 that is congruent mod 13 to $f(x) = x^6 - 1$ for $x = 0, 1, \dots, 12$. (b) Find $q(x)$ and r for which $x^6 - 1 \equiv (x-1)q(x) + r \bmod 13$. (c) Find the roots of $x^5 + x^4 + x^3 + x^2 + x + 1 \bmod 13$. (d) Describe the relation between the answers to parts (a), (b), and (c).

2. Write $x^{13} - x \bmod 13$ as a product of polynomials of the form $x+a$. (Actually, thanks to Fermat's theorem, you can do this without computation.)

3. Find a polynomial $f(x)$ of degree 3 for which there is only 1 solution of $f(a) \equiv 0 \bmod 7$. Find one for which there are *no* solutions. (Trial-and-error.)

Chapter 15

Tables of Indices mod p

Given a prime p and a primitive root g mod p, every positive number $n < p$ is congruent to g^i mod p for exactly one positive number i less than p. This number is called the **index**[1] of n mod p relative to the chosen primitive root g mod p. In dealing with specific problems mod p—for example, in solving quadratic equations mod p—it is useful to have tables giving the indices mod p of the positive numbers less than p and tables of the inverse function, giving the numbers when the index is known. Such tables were useful enough to students of number theory in the 19th century that the well-known mathematician C. G. J. Jacobi *published* tables of indices [**J**] for all primes less than 1000.

As was seen in Chapter 8, today it is easy with a programmable calculator to compute the inverse function—that is, to find g^i mod p when i is given—but even with a calculator indices are not easy. Tables of logarithms, to which Jacobi's tables of indices are strongly analogous, are utterly obsolete today because calculators have such functions built in, but Jacobi's tables are somewhat less obsolete. Of course a computer can easily be programmed to generate them, but

[1] In this definition, the index of 1 is $p - 1$, but one could obviously also consider 0 to be the index of 1 mod p. In any event, it is natural to regard indices as being numbers that are only determined mod $p - 1$, in which case it makes no difference whether 0 or $p - 1$ is given as the index of 1. Jacobi always lists the value $p - 1$.

the needed computations do not seem to lie within the practical range of programmable calculators.

In any case, whether such tables are useful or not, and whether calculators have made their construction obsolete or not, it is useful *conceptually* in dealing with the arithmetic of numbers mod p to think in terms of the index function (relative to a given primitive root) and its inverse, and a table is nothing but a very concrete presentation of that function.

Jacobi used a format like the following:

$p = 7$

numbers

1	*2*	*3*	*4*	*5*	*6*
3	2	6	4	5	1

indices

1	*2*	*3*	*4*	*5*	*6*
6	2	1	4	5	3

Each of these tables gives a function. The top rows, in italics, are to be regarded as captions and the numbers below them as values of the function for the numbers in the captions. The table on the left gives the first six powers of the primitive root 3 mod 7 and the table on the right gives the inverse function, which is the index; it tells for each number from 1 to 6 which power of the primitive root is congruent to that number mod 7.

The table on the left is easy to construct using a calculator, and then the one on the right can be found by interchanging the rows and rearranging the columns to put the top row in ascending order. When the number of numbers tabulated is greater than 9, Jacobi arranged the tables in the following format:

$p = 17$

numbers

	0	*1*	*2*	*3*	*4*	*5*	*6*	*7*	*8*	*9*
0		10	15	14	04	06	09	05	16	07
1	02	03	13	11	08	12	01			

indices

	0	*1*	*2*	*3*	*4*	*5*	*6*	*7*	*8*	*9*
0		16	10	11	04	07	05	09	14	06
1	01	13	15	12	03	02	08			

In these tables, the primitive root 10 mod 17 is used, as is seen from the fact that it is the value of the function for 1 in the numbers table. The placement of the numbers in the table describes, as the italicized row and column captions indicate, the values of the independent variable to which they belong. For example, the number 12 in the table of indices in the column captioned 3 and the row captioned 1 means that 12 is the index of 13. The steps in the construction of the tables are: First, choose a primitive root mod 17. (Jacobi always used 10 if, as in the case of 17, it was a primitive root.) Then construct the "numbers" table simply by computing the successive powers of the chosen primitive root. Finally, "invert" the numbers table to find the table of indices. For example, the entry in the column captioned 5 and the row captioned 1 of the table of indices is determined by locating the entry 15 in the numbers table; since it is in the column captioned 2 of the row captioned 0, the entry in the index table is 2.

Figure 1 reproduces Jacobi's tables for two larger primes, namely, for $p = 151$ and $p = 157$.

The quadratic formula reduces the solution of a congruence of the form $ax^2+bx+c \equiv 0 \bmod p$ to finding a square root of $b^2-4ac \bmod p$. But square roots mod p are easy to find using a table of indices by virtue of:

Proposition. *If the index i of a number mod p is odd, the number has no square root mod p. If i is even, then the number has two square roots mod p, namely, the numbers whose indices are $\frac{i}{2}$ and $\frac{i+p-1}{2}$.*

Proof. Let n be a given number less than p and let i be its index mod p. If $n \equiv m^2 \bmod p$, then the index of $m \bmod p$, call it j, satisfies $2j \equiv i \bmod (p-1)$, because $n \equiv m^2 \equiv g^{2j} \bmod p$. Because $p-1$ is even, this congruence has no solution x when i is odd, and it has the two solutions mod $(p-1)$ given in the statement of the proposition when i is even. □

Similarly, to find a cube root mod p of a number whose index mod p is i, one needs to find all solutions j of $3j \equiv i \bmod (p-1)$. The solution of this problem depends on whether $p-1$ is divisible by 3. If not, then 3 has a reciprocal mod $(p-1)$ and the congruence $3j \equiv i \bmod (p-1)$ has exactly one solution j for any given i. Thus, in this case, each number relatively prime to p has a unique cube root mod p. On the other hand, if 3 divides $p-1$, the congruence

$p = 151$

$p - 1 = 2 \cdot 3 \cdot 5^2$

Numeri.

I	0	1	2	3	4	5	6	7	8	9
		114	10	83	100	75	94	146	34	101
1	38	104	78	134	25	132	99	112	84	63
2	85	26	95	109	44	33	138	28	21	129
3	59	82	137	65	11	46	110	7	43	70
4	128	96	72	54	116	87	103	115	124	93
5	32	24	18	89	29	135	139	142	31	61
6	8	6	80	60	45	147	148	111	121	53
7	2	77	20	15	49	150	37	141	68	51
8	76	57	5	117	50	113	47	73	17	126
9	19	52	39	67	88	66	125	56	42	107
10	118	13	123	130	22	92	69	14	86	140
11	105	41	144	108	81	23	55	79	97	35
12	64	48	36	27	58	119	127	133	62	122
13	16	12	9	120	90	143	145	71	91	106
14	4	3	40	30	98	149	74	131	136	102
15	1									

Indices.

N	0	1	2	3	4	5	6	7	8	9
		150	70	141	140	82	61	37	60	132
1	2	34	131	101	107	73	130	88	52	90
2	72	28	104	115	51	14	21	123	27	54
3	143	58	50	25	8	119	122	76	10	92
4	142	111	98	38	24	64	35	86	121	74
5	84	79	91	69	43	116	97	81	124	30
6	63	59	128	19	120	33	95	93	78	106
7	39	137	42	87	146	5	80	71	12	117
8	62	114	31	3	18	20	108	45	94	53
9	134	138	105	49	6	22	41	118	144	16
10	4	9	149	46	11	110	139	99	113	23
11	36	67	17	85	1	47	44	83	100	125
12	133	68	129	102	48	96	89	126	40	29
13	103	147	15	127	13	55	148	32	26	56
14	109	77	57	135	112	136	7	65	66	145
15	75									

$p = 157$

$p - 1 = 2^2 \cdot 3 \cdot 13$

Numeri.

I	0	1	2	3	4	5	6	7	8	9
		139	10	134	100	84	58	55	109	79
1	148	5	67	50	42	29	106	133	118	74
2	81	112	25	21	93	53	145	59	37	119
3	56	91	89	125	105	151	108	97	138	28
4	124	123	141	131	154	54	127	69	14	62
5	140	149	144	77	27	142	113	7	31	70
6	153	72	117	92	71	135	82	94	35	155
7	36	137	46	114	146	41	47	96	156	18
8	147	23	57	73	99	102	48	78	9	152
9	90	107	115	128	51	24	39	83	76	45
10	132	136	64	104	12	98	120	38	101	66
11	68	32	52	6	49	60	19	129	33	34
12	16	26	3	103	30	88	143	95	17	8
13	13	80	130	15	44	150	126	87	4	85
14	40	65	86	22	75	63	122	2	121	20
15	111	43	11	116	110	61	1			

Indices.

N	0	1	2	3	4	5	6	7	8	9
		156	147	122	138	11	113	57	129	88
1	2	152	104	130	48	133	120	128	79	116
2	149	23	143	81	95	22	121	54	39	15
3	124	58	111	118	119	68	70	28	107	96
4	140	75	14	151	134	99	72	76	86	114
5	13	94	112	25	45	7	30	82	6	27
6	115	155	49	145	102	141	109	12	110	47
7	59	64	61	83	19	144	98	53	87	9
8	131	20	66	97	5	139	142	137	125	32
9	90	31	63	24	67	127	77	37	105	84
10	4	108	85	123	103	34	16	91	36	8
11	154	150	21	56	73	92	153	62	18	29
12	106	148	146	41	40	33	136	46	93	117
13	132	43	100	17	3	65	101	71	38	1
14	50	42	55	126	52	26	74	80	10	51
15	135	35	89	60	44	69	78			

Figure 1. Page 13 of Jacobi's *Canon Arithmeticus* [J], giving tables of numbers and indices for the primes 151 and 157.

$3j \equiv i \bmod (p-1)$ has no solutions if i is not divisible by 3 and has 3 solutions when i is divisible by 3, namely, $j = \frac{i}{3}$, $j = \frac{i+p-1}{3}$, and $j = \frac{i+2(p-1)}{3}$.

Thus, an index table for p makes it possible to find all cube roots mod p of a given number, and, in the same way, makes it possible to find all fourth roots, fifth roots, and so forth, mod p.

An index table also serves the function that was served by tables of logarithms in the pre-computer era, namely, the conversion of multiplication into addition. To multiply two numbers mod p, find their indices in the table of indices, add them, and use the table of numbers to find the number of which this sum is the index.

Exercises for Chapter 15

1. Jacobi's tables for the prime 31 are based on the primitive root 17 mod 31. Construct these tables.

2. Choose a prime p, find a primitive root mod p, and construct the corresponding tables.

Questions based on Jacobi's tables for $p = 151$ and $p = 157$:

3. Modulo one of the primes 151 and 157 the polynomial $x^2 + 2x - 1$ has two roots, and modulo the other it has none. Use Jacobi's tables to determine which is which, and find the roots in the case in which there are roots. Verify that $x^2+2x-1 \equiv (x-r_1)(x-r_2) \bmod p$, where p is the prime for which there are two roots and r_1 and r_2 are the roots.

4. Use Jacobi's table to find three primitive roots mod 157. How many primitive roots are there mod 157? How many mod 151?

5. Find a number that has no cube root mod 157. Find a number that does have a cube root mod 157, and find *all* of its cube roots. Verify by cubing on your calculator and checking that the cube minus the number is divisible by 157.

6. Find the order of 30 mod 157. Find the order of 90 mod 157.

7. Use the tables to compute a million (that is, 10^6) mod 151 and mod 157. (You can find both answers using mental calculation alone.) Check your answer using a calculator.

Chapter 16

Brahmagupta's Formula and Hypernumbers

The next several chapters are devoted to the solution of the problem $A\square + B = \square$ posed in Chapter 2. A treatise written in the 7th century by the Indian mathematician Brahmagupta contains an observation that is a key tool in the solution of this problem. Brahmagupta did not use symbolic algebra, so his observation is stated in words instead of symbols, which makes even his *statement* of the formula difficult to follow; that the formula was *discovered* and *proved* without the use of symbolic algebra is amazing. Very possibly, the formula is evidence of a mathematical culture that had been preserved and extended in various parts of the world ever since the Pythagorean and Greek works of a thousand years before (see Chapter 2).

Brahmagupta's Formula. *If A, B, B', x, y, u, and v are numbers for which*

$$Ax^2 + B = y^2$$

and

$$Au^2 + B' = v^2,$$

then

$$A(xv + yu)^2 + BB' = (Axu + yv)^2.$$

In other words, a solution of $A\square + B = \square$ can be combined with a solution of $A\square + B' = \square$ to find a solution of $A\square + BB' = \square$.

Proof. The proof using symbolic algebra is easy, but not trivial. Multiply the given equations to find

$$A^2x^2u^2 + Ax^2B' + Au^2B + BB' = y^2v^2$$

and add $A^2x^2u^2$ to both sides to find

$$(A^2x^2u^2 + Ax^2B') + (A^2x^2u^2 + Au^2B) + BB' = A^2x^2u^2 + y^2v^2,$$

which is to say

$$Ax^2 \cdot v^2 + Au^2 \cdot y^2 + BB' = A^2x^2u^2 + y^2v^2$$

from which Brahmagupta's formula

$$A(xv + yu)^2 + BB' = (Axu + yv)^2$$

follows when $2Axuyv$ is added to both sides. $\square$

In modern texts on number theory, Brahmagupta's formula is most often stated, "The norm of the product of $y+x\sqrt{A}$ and $v+u\sqrt{A}$ is the product of their norms." This statement is easily proved once meaning is given to the symbols $y+x\sqrt{A}$ and $v+u\sqrt{A}$ and to algebraic operations with them, which is normally done in an *ad hoc* way that asks the reader to accept without question that $x+y\sqrt{A}$ is a "number" when x and y are *integers* (possibly negative) and A is a number that is not a square.

Granted the free use of these symbols, Brahmagupta's formula follows when one multiplies the product

$$(y + x\sqrt{A})(v + u\sqrt{A}) = (yv + Axu) + (xv + yu)\sqrt{A} \tag{1}$$

and its "conjugate"

$$(y - x\sqrt{A})(v - u\sqrt{A}) = (yv + Axu) - (xv + yu)\sqrt{A}$$

to find

$$(y^2 - Ax^2)(v^2 - Au^2) = (yv + Axu)^2 - A(xv + yu)^2,$$

which is Brahmagupta's formula when B and B' are defined to be $y^2 - Ax^2$ and $v^2 - Au^2$, respectively. The convenience of this mnemonic for Brahmagupta's formula is the prime reason for dealing with what we will call *hypernumbers* in the following chapters.

We define a **hypernumber** for a given number A that is not a square[1] to be a formal expression of the form $y + x\sqrt{A}$, where x and y are numbers and where $\sqrt{A}$ is merely a symbol. Like numbers, hypernumbers can be added and multiplied. Addition of hypernumbers is defined by

$$(2) \qquad (y + x\sqrt{A}) + (v + u\sqrt{A}) = (y + v) + (x + u)\sqrt{A}$$

and their multiplication is defined by formula (1).

As is easily checked, the usual commutative, associative, and distributive laws of addition and multiplication hold for these operations, so computations with hypernumbers follow the same rules as computations with numbers, although the real meaning of such computations may cause the reader some philosophical unease. The usual attitude of mathematicians in such matters is to ignore philosophical issues and to regard the computations as a game that is played with clearly stated rules and to regard theorems about hypernumbers as statements about the possible outcomes of moves in the game.

In any discussion of hypernumbers, the value of A needs to be fixed at the outset. There will be no combining of hypernumbers with different values of A.

As an illustration of the power of computations with hypernumbers, note that in the case $A = 2$ the powers of the single hypernumber $1 + \sqrt{2}$ produce all the ratios in the Pythagorean sequence of Chapter 2. (See Exercise 1.) In the case $A = 3$, the powers of the single hypernumber $2 + \sqrt{3}$ produce all the solutions of $3x^2 + 1 = y^2$ found in Chapter 2, and the formula $(2 + \sqrt{3})^n(1 + \sqrt{3})$ for $n = 0, 1, 2, \ldots$ produces all the solutions of $3x^2 - 2 = y^2$ (see Exercise 2). For larger values of A and B, however, the use of hypernumbers in solving

[1] If A is a square, there is no need to regard $\sqrt{A}$ as a symbol. In that case, $\sqrt{A}$ is a number and the arithmetic of expressions $y + x\sqrt{A}$ is simply the arithmetic of numbers.

$A\Box + B = \Box$ becomes somewhat more difficult, as will be seen in the following chapters.

Exercises for Chapter 16

1. Describe the Pythagorean sequence $\frac{1}{1}, \frac{3}{2}, \frac{7}{5}, \frac{17}{12}, \frac{41}{29}, \frac{99}{70}, \frac{239}{169}, \frac{577}{408}, \ldots$ of Chapter 2 in terms of the powers of the hypernumber $1 + \sqrt{2}$.

2. The sequence of Exercise 2 in Chapter 2 can be seen as a combination of the two sequences $(2 + \sqrt{3})^n$ (the terms with even index) and $(1 + \sqrt{3})(2 + \sqrt{3})^n$ (the terms with odd index). Use this description to derive the formulas for generating the sequence that were found in that previous exercise.

3. Describe the sequence of Exercise 4, Chapter 2, in terms of hypernumbers.

Chapter 17

Modules of Hypernumbers

The notion of congruence of two numbers modulo two other numbers (see Chapter 4) extends immediately to hypernumbers for a given A. To say that two hypernumbers m and n are **congruent** modulo two other hypernumbers a and b (all for the same A) means that there are hypernumbers s, t, u, v for that A for which $m+sa+tb = n+ua+vb$. More generally, $m \equiv n \bmod [a_1, a_2, \ldots, a_k]$ for a list of hypernumbers $a_1, a_2, \ldots, a_k$ means that there are hypernumbers $r_1, r_2, \ldots, r_k$ and $s_1, s_2, \ldots, s_k$ for which $m + r_1a_1 + r_2a_2 + \cdots + r_ka_k = n + s_1a_1 + s_2a_2 + \cdots + s_ka_k$.

The statement that two lists $a_1, a_2, \ldots, a_k$ and $b_1, b_2, \ldots, b_l$ determine the same congruence relation in this way will be represented by the equation

$$[a_1, a_2, \ldots, a_k] = [b_1, b_2, \ldots, b_l].$$

In other words, this equation states symbolically that two hypernumbers m and n are congruent mod $[a_1, a_2, \ldots, a_k]$ if and only if they are congruent mod $[b_1, b_2, \ldots, b_l]$. A list of hypernumbers (for a given A) written between square brackets like this to indicate that the list is to be considered as a modulus in a congruence will be called a **module**, and when $[a_1, a_2, \ldots, a_k] = [b_1, b_2, \ldots, b_l]$, the two modules will be called **equal**.

In Chapter 4, where the modules were lists of numbers rather than hypernumbers, the comparison of modules was easy. The Euclidean algorithm "subtract the lesser from the greater" made it possible, using the rules $[a, 0] = [a]$ and $[a, b] = [a, b - a]$ (when $a \leq b$), to put any module $[a_1, a_2, \ldots, a_k]$ in the form

$$[a_1, a_2, \ldots, a_k] = [d]$$

for a single number d. Two modules of numbers $[d_1]$ and $[d_2]$ are equal if and only if $d_1 = d_2$, so this method solves the problem below in the case of numbers. In the case of hypernumbers the solution is slightly more difficult and much more interesting.

Problem. *Given two lists of hypernumbers $a_1, a_2, \ldots, a_k$ and $b_1, b_2, \ldots, b_l$, for a given A, determine whether $[a_1, a_2, \ldots, a_k] = [b_1, b_2, \ldots, b_l]$.*

Note that the very definition of equality of modules easily implies that it is reflexive, symmetric, and transitive, and that a module is equal to any module obtained by rearranging its terms. The complete solution of the problem, which will be given in Chapter 18, depends on the following rule for transforming a given module:

Theorem. *If $c \equiv 0 \bmod [a, b]$, then $[a, b, c] = [a, b]$. More generally, if $a_k \equiv 0 \bmod [a_1, a_2, \ldots, a_{k-1}]$, then $[a_1, a_2, \ldots, a_k] = [a_1, a_2, \ldots, a_{k-1}]$.*

Proof. For simplicity of notation, just the first case will be considered; the proof in the general case is essentially the same.

The assumption $c \equiv 0 \bmod [a, b]$ means that there are hypernumbers s, t, u, v for which $c + sa + tb = ua + vb$. What is to be shown is that then, for given hypernumbers m and n, there is an equation of the form

$$m + Sa + Tb = n + Ua + Vb \tag{1}$$

for some hypernumbers S, T, U, V if and only if there is an equation of the form

$$m + \mathcal{S}a + \mathcal{T}b + \mathcal{F}c = n + \mathcal{U}a + \mathcal{V}b + \mathcal{G}c \tag{2}$$

for some hypernumbers $\mathcal{S}$, $\mathcal{T}$, $\mathcal{F}$, $\mathcal{U}$, $\mathcal{V}$, $\mathcal{G}$. Of course (1) implies (2) when one sets $\mathcal{S} = S$, $\mathcal{T} = T$, $\mathcal{F} = 0$, $\mathcal{U} = U$, $\mathcal{V} = V$, and $\mathcal{G} = 0$. For the proof of the converse, let $(\mathcal{F} + \mathcal{G})(sa + tb)$ be added to both sides of (2) to find

$$m + \mathcal{S}a + \mathcal{T}b + \mathcal{F}(c + sa + tb) + \mathcal{G}(sa + tb) \\ = n + \mathcal{U}a + \mathcal{V}b + \mathcal{F}(sa + tb) + \mathcal{G}(c + as + tb).$$

The equation $c + sa + tb = ua + vb$ then makes it possible to eliminate c and conclude that (1) holds for $S = \mathcal{S} + \mathcal{F}u + \mathcal{G}s$, $T = \mathcal{T} + \mathcal{F}v + \mathcal{G}t$, $U = \mathcal{U} + \mathcal{F}s + \mathcal{G}u$, $V = \mathcal{V} + \mathcal{F}t + \mathcal{G}v$. □

Corollary 1. *If a and b are hypernumbers and if a can be subtracted from b, then $[a, b] = [a, b - a]$.*

Deduction. The equation $(b - a) + a = b$ implies both $b - a \equiv 0 \bmod [a, b]$ and $b \equiv 0 \bmod [a, b - a]$, so the theorem implies both $[a, b] = [a, b, b - a]$ and $[a, b - a] = [a, b - a, b]$, from which the desired conclusion follows by transitivity. □

Corollary 2. *If two modules are equal, then each can be transformed into the other by a sequence of steps in which the theorem is used simply to annex one entry to the list or drop one entry from the list.*

Deduction. Suppose $[a_1, a_2, \ldots, a_k] = [b_1, b_2, \ldots, b_l]$. Then each b is congruent to 0 mod $[a_1, a_2, \ldots, a_k]$ and each a is congruent to 0 mod $[b_1, b_2, \ldots, b_l]$. The theorem can therefore be used to annex $b_1, b_2, \ldots, b_l$ to $[a_1, a_2, \ldots, a_k]$ in l steps and conclude that $[a_1, a_2, \ldots, a_k] = [a_1, a_2, \ldots, a_k, b_1, b_2, \ldots, b_l]$. In the same way, then, $a_1, a_2, \ldots, a_k$ can be dropped in k steps to conclude that this module is also equal to $[b_1, b_2, \ldots, b_l]$. □

There is a natural way to *multiply* modules, namely, to define the product of the module $[a_1, a_2, \ldots, a_k]$ and the module $[b_1, b_2, \ldots, b_l]$ to be the module $[c_1, c_2, \ldots, c_{kl}]$ in which the c's consist of all kl products in which the first factor is chosen from $a_1, a_2, \ldots, a_k$ and the second factor is chosen from $b_1, b_2, \ldots, b_l$.

More precisely, this defines an operation on *lists* of hypernumbers. For it to be considered an operation on *modules,* the resulting module must be shown to be the same when $[a_1, a_2, \ldots, a_k]$ is replaced by another presentation of the same module. That is, one needs to prove:

Proposition. *If the product of two lists a and b is defined as above—the product $a \cdot b$ of a list a containing k hypernumbers and a list b containing l hypernumbers is a list obtained by putting the kl hypernumbers $a_i b_j$ in some order—and if c is another list of hypernumbers for which $[a] = [c]$, then $[a \cdot b] = [c \cdot b]$.*

Proof. If a is the list $a_1, a_2, \ldots, a_k$ and c is the list $c_1, c_2, \ldots, c_m$, Corollary 2 above states that the transition from $[a]$ to $[c]$ can be made by a sequence of steps in which one hypernumber in the list which is 0 mod the remaining hypernumbers in the list is either annexed to or dropped from the list. Therefore, the theorem will be proved if it is shown that $[a \cdot b] = [c \cdot b]$ for one such step—say for a step in which the list c is the list a with one hypernumber that is 0 mod $[a]$ annexed to it.

But in this case $c \cdot b$ is $a \cdot b$ with new terms annexed—specifically, with terms $c_1 b_j$ annexed, where c_1 is the sole entry in the list c that is not in the list a, and b_j is an entry in the list b—and all that needs to be shown is that the annexed terms are all 0 mod $[a \cdot b]$. But $c_1 b_j \equiv 0 \bmod [a \cdot b]$ follows from $c_1 \equiv 0 \bmod [a]$ when one multiplies an equation demonstrating $c_1 \equiv 0 \bmod [a]$ by b_j, which completes the proof. □

Multiplication of modules is of course commutative and associative because the multiplication of hypernumbers is commutative and associative.

Exercises for Chapter 17

1. Prove the following equations of modules:

(a) $[7 + 2\sqrt{11}] = [5, 7 + 2\sqrt{11}] = [5, 1 + \sqrt{11}]$,

(b) $[7 + \sqrt{7}, 13 + 4\sqrt{7}] = [3, 1 + \sqrt{7}]$. [Hint: Find *numbers* that are $\equiv 0$ for this modulus.]

2. The equation $[7 + 2\sqrt{11}][\sqrt{11}] = [5, 1 + \sqrt{11}][\sqrt{11}]$ implies that both $5\sqrt{11}$ and $11 + \sqrt{11}$ are 0 mod $[22 + 7\sqrt{11}]$. Prove these statements by finding explicit equations $5\sqrt{11} + a \cdot (22 + 7\sqrt{11}) = b \cdot (22 + 7\sqrt{11})$ and $11 + \sqrt{11} + c \cdot (22 + 7\sqrt{11}) = d \cdot (22 + 7\sqrt{11})$ in which a, b, c, and d are hypernumbers.

3. As will be proved in the next chapter, every module (for a given A) is equal to one in the form $[ef, eg + e\sqrt{A}]$ (unless it is $[0]$). Try to construct your own proof of this fact. Moreover, one can even stipulate that $g^2 \equiv A \bmod f$.

4. Provided f and F are relatively prime, a product module of the form $[f, g + \sqrt{A}][F, G + \sqrt{A}]$, in which $g^2 \equiv A \bmod f$ and $G^2 \equiv A \bmod F$, is equal to $[fF, \mathcal{G} + \sqrt{A}]$ where $\mathcal{G}$ is determined via the Chinese remainder theorem via $\mathcal{G} \equiv g \bmod f$ and $\mathcal{G} \equiv G \bmod F$. Prove this equation.

Chapter 18

A Canonical Form for Modules of Hypernumbers

Let a number A, not a square, be fixed, and let all hypernumbers be hypernumbers for that A.

Problem. *Given two lists* $a_1, a_2, \ldots, a_k$ *and* $b_1, b_2, \ldots, b_l$ *of hypernumbers, determine whether* $[a_1, a_2, \ldots, a_k] = [b_1, b_2, \ldots, b_l]$.

In the case of modules of numbers in Chapter 4 this problem was solved by showing that every module is equal to one in the form $[d]$ and that two in this form are equal only when they are *identical.* The problem in the case of hypernumbers can be solved in an analogous way by establishing a **canonical form** for such modules, a form with the property that every module is equal to one in this form, and two in this form are equal only when they are identical.

The congruence relation determined by a list in which all entries are zero is simply the relation of *equality,* but if a list contains even one nonzero entry, that entry is congruent to zero for the corresponding congruence relation without being equal to zero, so the congruence relation is different from equality. Thus, if $a_1, a_2, \ldots, a_k$ are all zero, $[a_1, a_2, \ldots, a_k] = [b_1, b_2, \ldots, b_l]$ if and only if $b_1, b_2, \ldots, b_l$ are all zero, and the solution of the full problem stated above is reduced to the solution of the case in which the given lists both contain nonzero entries. A canonical form for such modules can be found by the following sequence of simplifying steps.

(1) Let the given list be $a_1, a_2, \ldots, a_k$ where $a_i = y_i + x_i\sqrt{A}$, and let e be the greatest common divisor of the $2k$ coefficients $x_1, x_2, \ldots, x_k, y_1, y_2, \ldots, y_k$ of the entries. (This definition of e makes use of the assumption that at least one a_i is not zero.) Then $a_i = e \cdot b_i$ defines a hypernumber b_i, and the given module can be written in the form $[a_1, a_2, \ldots, a_k] = [e][b_1, b_2, \ldots, b_k]$, where the $2k$ coefficients of $b_1, b_2, \ldots, b_k$ have no common divisor greater than 1.

(2) Given a module $[b_1, b_2, \ldots, b_k]$ with this added property that the greatest common divisor of the coefficients of the b_i is 1, annex to the list, if it does not already contain one, a nonzero *number* that is 0 mod $[b_1, b_2, \ldots, b_k]$. Such a number is easy to find. For example, if $b = y + x\sqrt{A}$ is a nonzero entry in the list, then both $yb = y^2 + xy\sqrt{A}$ and $x\sqrt{A} \cdot b = xy\sqrt{A} + Ax^2$ are 0 mod $[b_1, b_2, \ldots, b_k]$, so $y^2 \equiv y^2 + xy\sqrt{A} + Ax^2 \equiv Ax^2 \bmod [b_1, b_2, \ldots, b_k]$. Therefore, the number $|y^2 - Ax^2|$ can be annexed to the list without changing the module. (Here $|y^2 - Ax^2|$ denotes, of course, the *difference* of y^2 and Ax^2, which is $y^2 - Ax^2$ if $y^2 \geq Ax^2$ and $Ax^2 - y^2$ otherwise. It is not zero, because Ax^2 is divisible an odd number of times by at least one of the prime divisors of A—because A is not a square—which would be impossible if Ax^2 were y^2.)

(3) Given a module $[b_1, b_2, \ldots, b_k]$ in which the greatest common divisor of the $2k$ coefficients $x_1, x_2, \ldots, x_k, y_1, y_2, \ldots, y_k$ of the k hypernumbers $b_i = y_i + x_i\sqrt{A}$ is 1, and in which b_1 is a nonzero number, annex to the list a hypernumber of the form $h + \sqrt{A}$ which is congruent to 0 mod $[b_1, b_2, \ldots, b_k]$. Such a hypernumber can be constructed in the following way. Because $1 \equiv 0 \bmod [x_1, x_2, \ldots, x_k, y_1, y_2, \ldots, y_k]$, there are numbers $r_1, r_2, \ldots, r_k, s_1, s_2, \ldots, s_k, u_1, u_2, \ldots, u_k, v_1, v_2, \ldots, v_k$ for which $1 + r_1x_1 + r_2x_2 + \cdots + r_kx_k + s_1y_1 + \cdots + s_ky_k = u_1x_1 + u_2x_2 + \cdots + u_kx_k + v_1y_1 + \cdots + v_ky_k$. Therefore, the coefficient of $\sqrt{A}$ in the hypernumber $r_1b_1 + r_2b_2 + \cdots + r_kb_k + s_1b_1\sqrt{A} + s_2b_2\sqrt{A} + \cdots + s_kb_k\sqrt{A}$ is 1 less than the coefficient of $\sqrt{A}$ in the hypernumber $u_1b_1 + u_2b_2 + \cdots + u_kb_k + v_1b_1\sqrt{A} + v_2b_2\sqrt{A} + \cdots + v_kb_k\sqrt{A}$, say the first is $Y_1 + X_1\sqrt{A}$ and the second is $Y_2 + X_2\sqrt{A}$

where $X_1+1 = X_2$. Then $Y_1+X_1\sqrt{A} \equiv 0 \equiv Y_2+(X_1+1)\sqrt{A} \bmod [b_1, b_2, \ldots, b_k]$. When q is a large enough number that Y_2+qb_1 is larger than Y_1, subtraction of $Y_1+X_1\sqrt{A}$ from $qb_1+Y_2+(X_1+1)\sqrt{A}$ gives a hypernumber that is 0 mod $[b_1, b_2, \ldots, b_k]$ in which the coefficient of $\sqrt{A}$ is 1, as required.

(4) A module $[b_1, b_2, \ldots, b_k]$ in which b_1 is a nonzero number and the coefficient of $\sqrt{A}$ in b_2 is 1 can be reduced to one of the form $[f, g+\sqrt{A}]$, in which $f \neq 0$ in the following way. If $k > 2$, each $b_i = y_i + x_i\sqrt{A}$ for $i > 2$ can be replaced by $b_i + qb_1 - x_i b_2$ when q is large enough to make the subtraction possible. When this replacement is made for each $i > 2$, the list takes the form b_1, $b_2 = y_2+\sqrt{A}$, n_3, n_4, $\ldots$, n_k where the n_i are *numbers* because the coefficients of $\sqrt{A}$ in b_i and $x_i b_2$ are the same. This module is, by the Euclidean algorithm, equal to $[f, y_2+\sqrt{A}]$ where f is the greatest common divisor of b_1, n_3, n_4, $\ldots$, n_k.

(5) Finally, a module of the form $[f, g+\sqrt{A}]$, where $f \neq 0$, can be assumed without loss of generality to have the properties that $g^2 \equiv A \bmod f$ and $g < f$. The first property can be assumed because if $g^2 \not\equiv A \bmod f$, one can annex the number $|g^2 - A|$ to the list (see step (2) above) and replace f and $|g^2 - A|$ in the resulting list with their greatest common divisor f' to find a representation $[f', g+\sqrt{A}]$ in which f' divides $|g^2 - A|$. As for the second property, if $g \geq f'$, one can simply subtract f' from g without changing the module and repeat until $g < f'$.

In summary:

Theorem. *Every module of hypernumbers for a given A, other than the trivial module $[0]$, is equal to one in the special form $[e][f, g+\sqrt{A}]$ where $ef \neq 0$, $g < f$, and $g^2 \equiv A \bmod f$.*

A module in this form $[e][f, g+\sqrt{A}]$—where $ef \neq 0$, $g < f$, and $g^2 \equiv A \bmod f$—will be said to be in **canonical form**. That two modules in this form are equal only if they are identical is proved in the corollary below.

Proposition. *Let* $[e][f, g+\sqrt{A}]$ *be a module in canonical form. Then a hypernumber* $Y + X\sqrt{A}$ *satisfies* $Y + X\sqrt{A} \equiv 0 \bmod [e][f, g + \sqrt{A}]$ *if and only if* $X \equiv 0 \bmod e$ *and* $Y \equiv gX \bmod ef$.

Proof. If $X \equiv 0 \bmod e$, say $X = eX'$, and $Y \equiv gX \bmod ef$, then $Y \equiv 0 \bmod e$, say $Y = eY'$; moreover, $Y' \equiv gX' \bmod f$, as follows directly from the meaning of $Y \equiv gX \bmod ef$, say $Y'+rf = gX'+sf$. Thus, $Y' + rf + X'\sqrt{A} = gX' + sf + X'\sqrt{A} = X'(g + \sqrt{A}) + sf$; multiply by e to find $Y + X\sqrt{A} + ref = X'(eg + e\sqrt{A}) + sef$, which shows $Y + X\sqrt{A} \equiv 0 \bmod [ef, eg + e\sqrt{A}]$, as required.

Conversely, if $Y+X\sqrt{A} \equiv 0 \bmod [ef, eg+e\sqrt{A}]$, then $X \equiv 0 \bmod e$ and $Y \equiv gX \bmod ef$, as can be shown in the following way. What is given is an equation $Y + X\sqrt{A} + a(ef) + b(eg + e\sqrt{A}) = c(ef) + d(eg + e\sqrt{A})$, where a, b, c, and d are hypernumbers. Since all terms other than $Y+X\sqrt{A}$ are hypernumbers in which both coefficients are divisible by e, the same is true of the hypernumber $Y + X\sqrt{A}$, say $Y+X\sqrt{A} = eY'+eX'\sqrt{A}$. In particular, $X \equiv 0 \bmod e$. Then division by e gives $Y' + X'\sqrt{A} + af + b(g + \sqrt{A}) = cf + d(g + \sqrt{A})$. Now the hypernumber $af+b(g+\sqrt{A})$ can be written in the form $jf+kf\sqrt{A}+l(g + \sqrt{A}) + m\sqrt{A}(g + \sqrt{A})$, where j, k, l, and m are *numbers,* and the same is true of the hypernumber $cf + d(g + \sqrt{A})$, so an equation of the form $Y' + X'\sqrt{A} + jf + kf\sqrt{A} + l(g + \sqrt{A}) + m(A + g\sqrt{A}) = j'f + k'f\sqrt{A} + l'(g + \sqrt{A}) + m'(A + g\sqrt{A})$ holds. The coefficients of $\sqrt{A}$ on the two sides are $X'+kf+l+mg = k'f+l'+m'g$, whereas the terms that are numbers are $Y'+jf+lg+mA = j'f+l'g+m'A$. When these equations are interpreted as congruences mod f and when use is made of $A \equiv g^2 \bmod f$, one finds first $X' + l + mg \equiv l' + m'g \bmod f$ and then $Y' + lg + mA \equiv l'g + m'A \equiv g(l' + m'g) \equiv g(X' + l + mg) \equiv gX' + lg + mA \bmod f$, from which the desired conclusion $Y' \equiv gX' \bmod f$—which is to say $Y \equiv gX \bmod ef$—follows. □

Corollary. *If* $[e][f, g+\sqrt{A}]$ *and* $[E][F, G+\sqrt{A}]$ *are modules in canonical form and if they are equal, then* $e = E$, $f = F$, *and* $g = G$.

Deduction. Let M denote the module $[e][f, g + \sqrt{A}] = [E][F, G + \sqrt{A}]$. Then $e = E$ because both are the largest number that divides all hypernumbers that are 0 mod M, $f = F$ because both ef and

$EF = eF$ are the smallest nonzero number that is $0 \bmod M$, and $g = G$ because both are the smallest number x for which $ex + e\sqrt{A}$ is $0 \bmod M$. □

Note that the canonical form of the module [1] is $[1, \sqrt{A}]$.

Exercises for Chapter 18

1. Find modules in canonical form equal to each of the following:
(a) $[7 + 2\sqrt{11}]$, (b) $[7, 2 + \sqrt{3}]$, (c) $[11, 10 + 2\sqrt{3}]$,
(d) $[25 + 6\sqrt{3}, 20 + 7\sqrt{3}]$.

2. Given numbers x and y, give rules for determining e, f, and g for which $[y + x\sqrt{A}] = [e][f, g + \sqrt{A}]$ and the module on the right is in canonical form.

3. For a given A, what condition must f satisfy in order for there to be a module $[f, g + \sqrt{A}]$ in canonical form?

4. Choose a value for A and write down two hypernumbers a and b for that A chosen more or less at random. Then put $[a, b]$ in canonical form. Most likely, you will find that the canonical form is $[a, b] = [1, \sqrt{A}]$. This is the analog for hypernumbers of the fact that two numbers chosen at random will most likely be relatively prime. Try to construct examples in which $[a, b] \neq [1, \sqrt{A}]$ but this fact is not obvious.

Chapter 19

Solution of $A\square + B = \square$

The method or methods by which ancient mathematicians found solutions of $A\square + B = \square$ have not survived. Archimedes' approximations $\frac{265}{153} < \sqrt{3} < \frac{1351}{780}$ strongly suggest that he had some general method, but we can only guess what it might have been. Brahmagupta in the 7th century gave a method by which the smallest solution $(x, y) = (120, 1151)$ of $92x^2 + 1 = y^2$ could be found, and the later Indian mathematician Bhāskara Achārya in the 12th century gave more general methods of solving $A\square + B = \square$.

The Indian methods were based on Brahmagupta's formula, employing it in an iterative way that was later called the "cyclic method" ([**D,** Chapter XII of Vol. 2] and [**E2,** Sec. 1.9]). The essence of the method can be expressed in the notation of modules of hypernumbers using:

Comparison Algorithm (so called because it gives a method of comparing two modules to determine whether they are equivalent; see Chapter 23).

Input: A module $[f, g + \sqrt{A}]$ in canonical form
Algorithm:
  r is the least solution of $r + g \equiv 0 \bmod f$ for which $r^2 > A$
  $F = \frac{r^2 - A}{f}$
  G is the least solution of $G \equiv r \bmod F$
Output: The module $[F, G + \sqrt{A}]$ in canonical form

The output module is in canonical form because the definitions imply $r^2 \equiv (-g)^2 \equiv A \bmod f$, $F \neq 0$, $G < F$, and $G^2 \equiv r^2 \equiv A \bmod F$. The input and output modules are related by the equation

$$(1) \qquad [r + \sqrt{A}][f, g + \sqrt{A}] = [f][F, G + \sqrt{A}],$$

where r is the number used by the algorithm to determine F and G, as can be proved as follows. First,

$$(2) \qquad r^2 + (g + \sqrt{A})(r + \sqrt{A}) = \frac{r+g}{f} \cdot f \cdot (r + \sqrt{A}) + A$$

because both sides are $r^2 + gr + A + g\sqrt{A} + r\sqrt{A}$. Because $\frac{r+g}{f}$ is a number, it follows that $r^2 \equiv A \bmod [f(r + \sqrt{A}), (g + \sqrt{A})(r + \sqrt{A})]$. Therefore, the number $r^2 - A = fF$ can be annexed to the list that describes $[r+\sqrt{A}][f, g+\sqrt{A}] = [f(r+\sqrt{A}), (g+\sqrt{A})(r+\sqrt{A})]$ to find that this module is $[f(r + \sqrt{A}), (r + \sqrt{A})(g + \sqrt{A}), fF]$. Moreover, when A is subtracted from both sides of (2), one finds $fF + (r + \sqrt{A})(g+\sqrt{A}) = \frac{r+g}{f} \cdot f \cdot (r+\sqrt{A})$, which means that $(r+\sqrt{A})(g+\sqrt{A})$ can be dropped from the list, leaving $[r + \sqrt{A}][f, g + \sqrt{A}] = [f \cdot (r + \sqrt{A}), fF] = [f][F, r + \sqrt{A}]$, which is (1), because $G \equiv r \bmod F$ by definition.

The use of the comparison algorithm to solve $A\Box + B = \Box$ will be explained below. The method that is explained finds **primitive** solutions of the problem, which is to say that it finds those solutions $Ax^2 + B = y^2$ in which x and y are relatively prime. If a solution is not primitive—say the greatest common divisor of x and y is $d > 1$—then d^2 must divide B, and $(u, v) = (\frac{x}{d}, \frac{y}{d})$ is a primitive solution of $Au^2 + \frac{B}{d^2} = v^2$. (In particular, *all* solutions are primitive unless B is divisible by a square greater than 1.) For this reason, it will suffice to have a method of constructing all *primitive* solutions. One merely needs to use such a method to construct the primitive solutions of $A\Box + \frac{B}{d^2} = \Box$ for each square factor d^2 of B (there may be no square factor greater than 1) and to use each one that is found (if any are) to construct a solution of $A\Box + B = \Box$.

All primitive solutions of a given equation $A\Box + B = \Box$ can be found in the following way.

1. Given A and B (A not a square, B not zero), find all square roots of A mod B. In the cases of greatest interest, B is a small number and this step can easily be accomplished by trial-and-error. (The process of finding square roots of A mod B is simplified by factoring B, finding square roots of A mod the factors, and using the Chinese Remainder Theorem to put them together.)

2. For each square root g of A mod B, apply the comparison algorithm repeatedly to the module $[B, g+\sqrt{A}]$ to generate a sequence of modules $[B, g+\sqrt{A}] = [f_0, g_0+\sqrt{A}]$, $[f_1, g_1+\sqrt{A}]$, $[f_2, g_2+\sqrt{A}]$, ... Each pair of successive modules satisfies a relation

$$[r_i+\sqrt{A}][f_{i-1}, g_{i-1}+\sqrt{A}] = [f_{i-1}][f_i, g_i+\sqrt{A}]. \tag{3}$$

As will be found (it will be proved in the next chapter that it must always happen) this sequence eventually begins to repeat. Therefore, it will always be possible to determine whether the sequence contains the module [1] and, if so, exactly which indices i satisfy $[f_i, g_i+\sqrt{A}] = [1]$, or, what is the same, satisfy $f_i = 1$. (As will be shown, if $f_i = 1$ ever occurs, it occurs infinitely often.)

3. For each square root g of A mod B and for each index i for which $f_i = 1$ in the sequence $[B, g+\sqrt{A}] = [f_0, g_0+\sqrt{A}]$, $[f_1, g_1+\sqrt{A}]$, $[f_2, g_2+\sqrt{A}]$, ... that follows from it, a primitive solution $Ax^2 + B = y^2$ of the given equation can be found in the following way. String together the n equations (3) that relate the successive modules $[f_i, g_i+\sqrt{A}]$ to find the equation

$$\begin{aligned}[r_1+\sqrt{A}][r_2+\sqrt{A}]\cdots[r_n+\sqrt{A}]&[B, g+\sqrt{A}] \\ &= [B][f_1][f_2]\ldots[f_{n-1}][f_n, g_n+\sqrt{A}]\end{aligned}$$

relating the initial module $[B, g+\sqrt{A}] = [f_0, g_0+\sqrt{A}]$ and the nth module $[f_n, g_n+\sqrt{A}] = [1]$. If one sets $Y + X\sqrt{A} = (r_1+\sqrt{A})(r_2+\sqrt{A})\cdots(r_n+\sqrt{A})$, this equation takes the form

$$[Y + X\sqrt{A}][B, g+\sqrt{A}] = [Bf_1f_2\cdots f_{n-1}].$$

In particular, the hypernumber $(Y + X\sqrt{A})B$ is divisible by the number $Bf_1f_2\cdots f_{n-1}$, which means that $Y + X\sqrt{A}$ is divisible by $f_1f_2\cdots f_{n-1}$, say $Y + X\sqrt{A} = (y + x\sqrt{A})f_1f_2\cdots f_{n-1}$. With x and

y defined in this way, the equation

$$[y + x\sqrt{A}][B, g + \sqrt{A}] = [B]$$

holds. As will be shown in the next chapter:

Theorem. *For each square root g of $A \bmod B$, one can determine quite explicitly the numbers n for which $f_n = 1$ in the nth module generated by the comparison algorithm when one begins with $[B, g + \sqrt{A}]$ and applies the algorithm iteratively. For each such g and n (if there are any), the pair of numbers (x, y) given by the above formula*

$$(4) \qquad y + x\sqrt{A} = \frac{(r_1 + \sqrt{A})(r_2 + \sqrt{A}) \cdots (r_n + \sqrt{A})}{f_1 f_2 \cdots f_{n-1}}$$

is a primitive solution of $Ax^2 + B = y^2$, and all *primitive solutions are found in this way.*

Example: In Brahmagupta's equation $92\square + 1 = \square$ mentioned at the beginning of this chapter, $B = 1$ and the only module $[B, g + \sqrt{92}]$ to be considered is $[1]$. For this input module, one finds that the sequence of r's is 10, 14, 12, 12, 14, 10, after which the sequence continues cyclically as 10, 14, 12, 12, ... , and the sequence of f's cycles through 1, 8, 13, 4, 13, 8, 1, 8, 13, 4, ... Thus the first solution (disregarding the trivial solution $92 \cdot 0^2 + 1 = 1^2$ corresponding to the initial 1) is

$$\frac{(10 + \sqrt{92})(14 + \sqrt{92})(12 + \sqrt{92})(12 + \sqrt{92})(14 + \sqrt{92})(10 + \sqrt{92})}{8 \cdot 13 \cdot 4 \cdot 13 \cdot 8}$$

which is easily found to be $1151 + 120\sqrt{92}$, the solution of Brahmagupta that was mentioned above. There are infinitely many solutions, one for each occurrence of $f = 1$, which happens at every sixth step. These later solutions are clearly the *powers* of the solution found above, so they are the coefficients of the hypernumbers $(1151+120\sqrt{92})^n$ for $n = 2, 3, 4, \ldots$. Note how quickly the numbers x and y in these solutions grow; when $n = 2$, $(x, y) = (276240, 2649601)$.

How to Organize the Computations

Write the sequences g, r_1, r_2, r_3, ... and B, f_1, f_2, f_3, ... on successive lines, with B between g and f_1, r_1 between B and f_1, f_2 between r_1 and r_2, and so forth. The computations of Brahmagupta's example are then displayed in the form

$(A = 92)$

```
0   10   14   12   12   14   10   10   14   12   12   14 ...
  1    8    13   4    13   8    1    8    13   4    13   8 ....
```

Each new term r_i on the top line is the least solution of $r_{i-1} + r_i \equiv 0 \bmod f_{i-1}$ for which $r_i^2 > A$—in particular, the sum of two consecutive numbers on the top line is always divisible by the number on the bottom line that lies between them—and each new f_i is given by the formula $f_i = \frac{r_i^2 - A}{f_{i-1}}$, so the successive terms are easy to calculate. The indices n for which $f_n = 1$ are easy to see, and the numbers r_i and f_i needed in formula (4) can be read off. (Note that the first number in the top row is g, the first number in the bottom row is B, and these numbers do not appear in formula (4).)

In organizing the reduction of the hypernumber given by formula (4), it is often easiest to cancel factors from the numerator and denominator during the course of computing the product in the numerator. For example, the numerator of the smallest solution of Brahmagupta's problem is $\big((10+\sqrt{92})(14+\sqrt{92})\big)^2(12+\sqrt{92})(12+\sqrt{92}) = \big(140+92+(10+14)\sqrt{92}\big)^2(144+92+24\sqrt{92}) = 8^2(29+3\sqrt{92})^2 \cdot 4 \cdot (59+6\sqrt{92})$. The factors $8^2 \cdot 4$ cancel from the denominator, leaving just 13^2. Then $\big((29+3\sqrt{92})(59+6\sqrt{92})\big)(29+3\sqrt{92}) = 13(259+27\sqrt{92})(29+3\sqrt{92}) = 13^2 \cdot (1151 + 120\sqrt{92})$ yields the answer given above.

Computation with Hypernumbers

A simple way to do machine computation with hypernumbers without having to write special programs is to treat $y + x\sqrt{A}$ as a 2×2 matrix of numbers $\begin{bmatrix} y & x \\ xA & y \end{bmatrix}$. Since the square of $\begin{bmatrix} 0 & 1 \\ A & 0 \end{bmatrix}$ is $\begin{bmatrix} A & 0 \\ 0 & A \end{bmatrix}$, the matrix corresponding to $(y_1 + x_1\sqrt{A})(y_2 + x_2\sqrt{A})$ is then the product of the matrix corresponding to $y_1+x_1\sqrt{A}$ and the matrix corresponding to $y_2+x_2\sqrt{A}$. Since the same is obviously true of sums, one can in this way do hypernumber computations using a computer software package or even a programmable calculator that handles matrices. In implementing formula (4), the numbers can be kept small by alternating multiplications and divisions. Specifically, multiply the first two factors in the numerator, divide by the first factor in the denominator, multiply by the next factor in the numerator, divide by the next factor in the denominator, and so forth; as is easy to prove, each division goes evenly, so the result at each step is a hypernumber.

Exercises for Chapter 19

1. Find all solutions of $79x^2+21=y^2$.

2. Find all solutions of $13x^2+1=y^2$.

3. Choose a number A, not a square, and a number x for which Ax^2 is slightly less than a square, say $Ax^2+B=y^2$ where B is not a very large number. Use the method of the chapter to find all solutions of $A\square+B=\square$. (Remember that the method produces only *primitive* solutions, so if your B has square factors $d^2>1$ you also have to look for solutions of $A\square+\frac{B}{d^2}=\square$.)

4. Find all solutions of $61\square+1=\square$. [Persevere. The smallest squares for which this equation holds are enormous, but not too difficult to find if the computation is organized carefully.]

5. The equation $A\square+1=\square$—i.e., the special case $B=1$ of the problem of this chapter—is called **Pell's equation**. Show that if (X,Y) is a solution of $AX^2+B=Y^2$ and if $Ax^2+1=y^2$ is the smallest solution of Pell's equation for this A, then the next solution of $AX^2+B=Y^2$ in the sequence that contains the solution (X,Y) is given by the coefficients of $(Y+X\sqrt{A})(y+x\sqrt{A})$.

6. Find a published table of solutions of Pell's equation and derive some of the entries using the method of this chapter. Fermat stated that Pell's equation has infinitely many solutions whenever A is not a square, a statement which follows from the method of this chapter and the fact, proved in Chapter 22, that [1] is stable.

Chapter 20

Proof of the Theorem of Chapter 19

For given numbers A and B (A not a square and B not zero), the theorem of Chapter 19 describes an algorithm for finding all primitive solutions of $Ax^2 + B = y^2$; it is based on finding all occurrences of $[1]$ in the sequence of modules $[B, g + \sqrt{A}] = [f_0, g_0 + \sqrt{A}]$, $[f_1, g_1 + \sqrt{A}]$, $[f_2, g_2 + \sqrt{A}]$, ... generated by applying the comparison algorithm iteratively to modules $[B, g + \sqrt{A}]$ in canonical form. (In particular, there are no primitive solutions if A is not a square mod B.)

Theorem 1. *The sequence of modules* $[f_0, g_0 + \sqrt{A}]$, $[f_1, g_1 + \sqrt{A}]$, $[f_2, g_2 + \sqrt{A}]$, ... *that results when the comparison algorithm is applied repeatedly to any module* $[f_0, g_0 + \sqrt{A}]$ *in canonical form eventually begins to repeat.*

Once a module in the sequence is repeated, all subsequent modules are repeats, so the infinite sequence contains only a finite number of *distinct* modules and eventually becomes an infinitely repeating cycle.

Proof. Let r_1, r_2, ... be the numbers used by the comparison algorithm to go from $[f_0, g_0 + \sqrt{A}]$ to $[f_1, g_1 + \sqrt{A}]$, from $[f_1, g_1 + \sqrt{A}]$ to $[f_2, g_2 + \sqrt{A}]$, and so forth. The key idea of the proof is that $|r_{i+1} - f_i|^2$ is less than $|r_i - f_{i-1}|^2$ unless $|r_i - f_{i-1}|^2 < A$; once an i is reached for which $|r_i - f_{i-1}|^2 < A$ (which may be true for $i = 1$), the same inequality holds for all subsequent values of i, which implies, as will be shown, that the sequence must begin to repeat.

First, $r_i + r_{i+1} \geq 2f_i$ for each i for the following reasons. Because $r_i + r_{i+1} \equiv g_i + r_{i+1} \equiv 0 \bmod f_i$ (the first by the definition of g_i, the second by the definition of r_{i+1}), $r_i + r_{i+1}$ is divisible by f_i, so $r_i + r_{i+1} \geq 2f_i$ will follow if $r_i + r_{i+1} > f_i$ is proved to hold. If $f_{i-1} \geq r_i$, then $r_i^2 > r_i^2 - A = f_i f_{i-1} \geq f_i r_i$, which implies $r_i > f_i$ and therefore implies the desired inequality $r_i + r_{i+1} > f_i$. Thus, the desired inequality holds unless $f_{i-1} < r_i \leq f_i$. But it holds when r_i is in this range too, because then $(r_i - f_{i-1})^2 < A$ (by the definition of r_i, because $r_i - f_{i-1}$ is less than r_i and congruent to it mod f_{i-1}), from which it follows that

$$
\begin{aligned}
r_i^2 - 2r_i f_{i-1} + f_{i-1}^2 &< A \\
f_{i-1} f_i + f_{i-1}^2 &< 2r_i f_{i-1} && \text{(add } 2r_i f_{i-1} \text{ and subtract } A\text{)} \\
f_i + f_{i-1} &< 2r_i && \text{(divide by } f_{i-1}\text{)}.
\end{aligned}
$$

Multiplication of the last inequality by f_i puts it in the form $f_i^2 + r_i^2 - A < 2r_i f_i$, which is the same as $(f_i - r_i)^2 < A$ because $r_i \leq f_i$. Since $A < r_{i+1}^2$ by the definition of r_{i+1}, this gives $(f_i - r_i)^2 < r_{i+1}^2$, from which the desired conclusion $f_i - r_i < r_{i+1}$ follows.

Next, $|r_i - f_{i-1}|^2 > A$ implies $f_{i-1} + r_{i+1} > f_i + r_i$ for the following reasons. (Here $|a-b|^2$ means, of course, $(a-b)^2$ if $a \geq b$ and otherwise means $(b-a)^2$. In either case, this number is $a^2 + b^2 - 2ab$.) In fact, $|r_i - f_{i-1}|^2 > A$ means $r_i^2 + f_{i-1}^2 > A + 2r_i f_{i-1}$. Since $r_i^2 - A = f_i f_{i-1}$, subtracting A from both sides and dividing by f_{i-1} gives $f_i + f_{i-1} > 2r_i$. Addition of this to the inequality $r_i + r_{i+1} \geq 2f_i$ of the last paragraph then gives $f_i + f_{i-1} + r_i + r_{i+1} > 2r_i + 2f_i$, which is the desired conclusion, i.e., $f_{i-1} + r_{i+1} > f_i + r_i$.

It can now be shown that $|r_i - f_{i-1}|^2 > A$ implies $|r_i - f_{i-1}|^2 > |r_{i+1} - f_i|^2$, as was claimed above. The proof of this implication will be divided into two cases, $f_i \leq r_{i+1}$ and $f_i > r_{i+1}$. In the first case, the definition of r_{i+1} implies that $|r_{i+1} - f_i|^2 < A$ so of course $|r_{i+1} - f_i|^2 < |r_i - f_{i-1}|^2$, as was to be shown. In the second case, the assumption $|r_i - f_{i-1}|^2 > A$ implies $f_{i-1} + r_{i+1} > f_i + r_i$, as was just shown, so $f_i > r_{i+1}$ implies $0 < f_i - r_{i+1} = f_i - r_{i+1} + r_i - r_i < f_{i-1} - r_i$, and $(f_i - r_{i+1})^2 < (f_{i-1} - r_i)^2$, as was to be shown.

Once an i is reached for which $|r_i - f_{i-1}|^2 < A$—as has now been proved must eventually happen—the same inequality holds for all subsequent values of i, as can be seen in the following way. The assumption $|r_i - f_{i-1}|^2 < A$ implies $f_i + f_{i-1} < 2r_i$ in the same way that $|r_i - f_{i-1}|^2 > A$ implies $f_i + f_{i-1} > 2r_i$ above. Therefore, this assumption implies $f_i^2 + f_{i-1}f_i < 2r_i f_i$, which is the same as $f_i^2 + r_i^2 - A < 2r_i f_i$ or $|f_i - r_i|^2 < A$. If $|r_{i+1} - f_i|^2$ were greater than A, then, by the definition of r_{i+1}, f_i would have to be greater than r_{i+1}. (Otherwise, $r_{i+1} - f_i$ would be a smaller solution of $r + g_i \equiv 0 \bmod f_i$ whose square was larger than A.) Since $r_i + r_{i+1} \geq 2f_i$, this would mean that $r_i > f_i$ and in fact that $r_i - f_i \geq f_i - r_{i+1}$, which would imply $(r_i - f_i)^2 \geq (f_i - r_{i+1})^2 > A$, which is not the case. Therefore, $|r_{i+1} - f_i|^2$ must be less than A, as was to be shown. (It cannot be equal to A because A is not a square.)

Therefore, from some point on, say for $i \geq N$, the inequality $|r_i - f_{i-1}|^2 < A$ must always be satisfied. The number $A - |r_i - f_{i-1}|^2 = A - r_i^2 - f_{i-1}^2 + 2r_i f_{i-1}$ is zero mod f_{i-1}. In other words, f_{i-1} divides one of the numbers A, $A-1$, $A-4$, $A-9$, ... (a terminating sequence). This observation limits the possible values of f_{i-1} for $i \geq N$ to a finite set. Since $g_{i-1} < f_{i-1}$, the values of f_{i-1} and g_{i-1} in $[f_{i-1}, g_{i-1} + \sqrt{A}]$ for $i \geq N$ are limited to a finite set and the infinite sequence must eventually contain a repeat, as was to be shown. □

After enough terms of the sequence $[B, g + \sqrt{A}] = [f_0, g_0 + \sqrt{A}]$, $[f_1, g_1 + \sqrt{A}]$, $[f_2, g_2 + \sqrt{A}]$, ... are computed to find the first repeat, the entire sequence is known and the occurrences, if any, of [1] in the sequence can be explicitly determined.

Theorem 2. *If n is an index for which $[f_n, g_n + \sqrt{A}] = [1]$, then the coefficients of the hypernumber*

$$y + x\sqrt{A} = \frac{(r_1 + \sqrt{A})(r_2 + \sqrt{A}) \cdots (r_n + \sqrt{A})}{f_1 f_2 \cdots f_{n-1}}$$

of formula (4) *in Chapter 19 are a primitive solution of $Ax^2 + B = y^2$.*

The proof of this theorem will make use of the concept of the **norm** of a module, which is the product of the module with its conjugate. By definition, the **conjugate** of a module in canonical form $[e][f, g + \sqrt{A}]$ is the module $[e][f, g' + \sqrt{A}]$ where g is the smallest solution of $g' + g \equiv 0 \bmod f$. Otherwise stated, $g' = f - g$ if $g > 0$, and $g' = 0$ if $g = 0$. Another way to describe the conjugate of a module is to say that if the necessary and sufficient conditions for $y + x\sqrt{A}$ to be divisible by the module are $x \equiv 0 \bmod e$ and $y \equiv gx \bmod ef$, where g is a square root of $A \bmod f$, then the necessary and sufficient conditions for $y + x\sqrt{A}$ to be divisible by the conjugate of the module are $x \equiv 0 \bmod e$ and $y + gx \equiv 0 \bmod ef$.

Lemma. *The norm of a module in canonical form $[e][f, g + \sqrt{A}]$, defined in this way to be the product of the module with its conjugate, is given by the explicit formula $[e^2 f][d, g + \sqrt{A}]$, where d is the greatest common divisor of f, $2g$, and $\frac{|A-g^2|}{f}$.*

The greatest common divisor of f, $2g$, and $\frac{|A-g^2|}{f}$ will be called the **content** of $[e][f, g + \sqrt{A}]$. If the content of a module is 1, the module will be called **primitive**.

Proof of the lemma. Given $[e][f, g + \sqrt{A}]$ in canonical form, let g' be the least solution of $g' + g \equiv 0 \bmod f$. It is to be shown that

$$[e][f, g + \sqrt{A}][e][f, g' + \sqrt{A}] = [e^2 f][d, g + \sqrt{A}]$$

where d is the content of $[e][f, g + \sqrt{A}]$. When the module on the left is multiplied out and e^2 is canceled from all terms on both sides, the equation to be proved becomes

$$[f^2, f(g + \sqrt{A}), f(g' + \sqrt{A}), gg' + A + (g + g')\sqrt{A}] = [f][d, g + \sqrt{A}].$$

For sufficiently large numbers s and t, the module on the left can be written $[f^2, fg + f\sqrt{A}, sf^2 + fg' - fg, tf^2 + gg' + A - \frac{g+g'}{f} \cdot fg] = [f^2, fg + f\sqrt{A}, sf^2 + fg' - fg, tf^2 + A - g^2]$, which is $[f][d_0, g + \sqrt{A}]$ when d_0 is defined by $[d_0] = [f, sf + g' - g, tf + \frac{A-g^2}{f}]$. This equation implies $f \equiv 0 \bmod d_0$, $g \equiv sf + g' \equiv g' \equiv -g \bmod d_0$, and $\frac{|A-g^2|}{f} \equiv 0 \bmod d_0$, so d_0 divides d. Conversely, f, $sf+g'-g$ and $tf+\frac{A-g^2}{f}$ are all divisible by d, so $d_0 = d$ and the proof is complete. □

Proof of Theorem 2. The content of $[f_{i-1}, g_{i-1} + \sqrt{A}]$ divides the content of $[f_i, g_i + \sqrt{A}]$, as one can see in the following way.

Let d be the content of $[f_{i-1}, g_{i-1} + \sqrt{A}]$. Because $r_i = uf_{i-1} - g_{i-1}$ for some u, the number $f_i f_{i-1} = r_i^2 - A$ can be written as $u^2 f_{i-1}^2 - 2uf_{i-1}g_{i-1} + g_{i-1}^2 - A$, from which it follows that df_{i-1} divides $f_i f_{i-1}$, so d divides f_i. Because $2r_i \equiv 2g_i \bmod f_i$ and therefore $2g_i \equiv 2r_i \bmod d$, it also follows that $2g_i \equiv 2uf_{i-1} - 2g_{i-1} \equiv 0 \bmod d$. Finally, $g_i \equiv r_i \bmod f_i$ means that $g_i + kf_i = r_i + lf_i$ for some numbers k and l; when both sides are squared and r_i^2 is replaced with $A + f_i f_{i-1}$, one finds an equation in which all terms are divisible by df_i except for the term g_i^2 in the first expression and the term A in the last. Therefore, $g_i^2 \equiv A \bmod df_i$ and d divides the content of $[f_i, g_i + \sqrt{A}]$, as was to be shown.

Therefore, $f_n = 1$ implies that $[f_0, g_0 + \sqrt{A}]$ is primitive and therefore implies, by the lemma, that $[B, g + \sqrt{A}][B, g' + \sqrt{A}] = [B]$ where $[B, g' + \sqrt{A}]$ is the canonical form of the conjugate of $[B, g + \sqrt{A}]$.

Multiply the equation

$$[y + x\sqrt{A}][B, g + \sqrt{A}] = [B]$$

that was found in Chapter 19 by $[B, g' + \sqrt{A}]$ and cancel $[B]$ to find

$$[y + x\sqrt{A}] = [B, g' + \sqrt{A}]$$

which gives the canonical form of $[y + x\sqrt{A}]$ and shows in particular that x and y are relatively prime. Application of Brahmagupta's formula to $(r_1 + \sqrt{A})(r_2 + \sqrt{A}) \cdots (r_n + \sqrt{A}) = Y + X\sqrt{A}$ gives $(r_1^2 - A)(r_2^2 - A) \cdots (r_n^2 - A) = Y^2 - AX^2$ and therefore shows that $Y^2 > AX^2$, and therefore, since $Y + X\sqrt{A}$ is the numerator of $y + x\sqrt{A}$, $y^2 > Ax^2$. The number $y^2 - Ax^2$, call it C, satisfies $[y + x\sqrt{A}] = [C, y + x\sqrt{A}]$ (see step (2) in Chapter 18). A common factor of C and x would be a factor of y^2, contrary to the fact that x and y are relatively prime. Therefore x has a reciprocal mod C, call it u, and $u(y + x\sqrt{A})$ can be annexed to the list to find $[y + x\sqrt{A}] = [C, y +$

$x\sqrt{A}, uy + ux\sqrt{A}]$. The middle term is congruent mod C to x times the last, so it can be dropped, leaving $[y + x\sqrt{A}] = [C, uy + ux\sqrt{A}]$, which is $[C, h + \sqrt{A}]$, where h is the least solution of $h \equiv uy \bmod C$. Since $h^2 \equiv u^2y^2 \equiv u^2Ax^2 \equiv A \bmod C$, this is the canonical form of $[y+x\sqrt{A}]$, so $[B, g'+\sqrt{A}]$ and $[C, h+\sqrt{A}]$ are identical. In particular, $B = C = y^2 - Ax^2$, as was to be shown. □

It remains to show that all solutions of $Ax^2+B = y^2$ are obtained in this way. That is:

Theorem 3. *Every primitive solution of $Ax^2 + B = y^2$ is found by the method of Theorem 2 for some square root g of $A \bmod B$ and for some index n.*

Proof. The proof will use the following algorithm to *reverse* the comparison algorithm.

Reduction Algorithm (so called because it reduces the value of X in the input equation).

Let $[B, g + \sqrt{A}]$ be a given module in canonical form.

Input: A hypernumber $Y + X\sqrt{A}$ in which $Y^2 > AX^2$ and $X > 0$ and for which the canonical form of the product $[Y + X\sqrt{A}][B, g + \sqrt{A}]$ is $[B][F, G + \sqrt{A}]$ for some F and G.

Algorithm:

Let s be the smallest solution of $s \equiv G \bmod F$ for which $sX \geq Y$.

Then $(s - \sqrt{A})(Y + X\sqrt{A})$ is defined and congruent to 0 mod F, say it is $F(Y_1 + X_1\sqrt{A})$.

Output: A hypernumber $Y_1 + X_1\sqrt{A}$ in which $Y_1^2 > AX_1^2$ (but X_1 may be zero) and the canonical form of the product $[Y_1 + X_1\sqrt{A}][B, g + \sqrt{A}]$ is $[B][F_1, G_1 + \sqrt{A}]$, where $F_1 = \frac{s^2-A}{F}$ and $G_1 + s \equiv 0 \bmod F_1$.

The hypernumber $(s-\sqrt{A})(Y+X\sqrt{A})$ is meaningful, which is to say that $\sqrt{A}(Y+X\sqrt{A})$ can be subtracted from $s(Y+X\sqrt{A})$. In the case of the coefficient of $\sqrt{A}$ this is simply the condition $sX \geq Y$ of the definition of s; in the case of the other coefficient, it follows from the observation that $(sX)^2 \geq Y^2 > AX^2$ and $X \neq 0$ imply $s^2 > A$, so $s^2Y^2 > A \cdot AX^2 = (AX)^2$ and $sY > AX$.

The input equation implies $(Y + X\sqrt{A})B \equiv 0 \bmod [BF, B(G + \sqrt{A})]$ and therefore implies $Y + X\sqrt{A} \equiv 0 \bmod [F, G + \sqrt{A}]$, directly from the definitions. By the corollary of Chapter 18, then,

$Y \equiv GX \bmod F$. Therefore, $Y \equiv sX \bmod F$, which shows that the coefficient of $\sqrt{A}$ in $(s - \sqrt{A})(Y + X\sqrt{A})$ is divisible by F. That the other coefficient is also divisible by F follows from the observation that it is $sY - AX \equiv GY - G^2X \equiv G(Y - GX) \equiv 0 \bmod F$. In short, the equation $(s-\sqrt{A})(Y+X\sqrt{A}) = F(Y_1+X_1\sqrt{A})$ determines $Y_1 + X_1\sqrt{A}$.

Because s is the *smallest* number congruent to G mod F for which $sX \geq Y$, the inequality $sX < FX+Y$ must hold, so $FX_1 = sX-Y < FX$ and $X_1 < X$.

That $AX_1^2 < Y_1^2$ follows from a simple variation of Brahmagupta's formula that is easy to prove (see Exercise 2), namely, the formula $A(sX - Y)^2 + (s^2 - A)(Y^2 - AX^2) = (sY - AX)^2$, which implies $AF^2X_1^2 < F^2Y_1^2$.

The output equation can be deduced in the following way. Because $s^2 \equiv G^2 \equiv A \bmod F$, the number $s^2 - A$ is divisible by F, say $s^2 - A = F_1F$. Let G_1 be the least solution of $G_1 + s \equiv 0 \bmod F_1$. The equation

$$\begin{aligned}(G_1 + \sqrt{A})(s + \sqrt{A}) + F_1F &= G_1s + (G_1 + s)\sqrt{A} + A + s^2 - A \\ &= (G_1 + s)(s + \sqrt{A}) = qF_1(s + \sqrt{A}),\end{aligned}$$

where q is defined by $G_1 + s = qF_1$, shows that the module $[(G_1 + \sqrt{A})(s+\sqrt{A}), F_1F, F_1(s+\sqrt{A})]$ is unchanged if either of the first two terms is dropped from the list. Thus

$$[F_1F, F_1(s + \sqrt{A})] = [(G_1 + \sqrt{A})(s + \sqrt{A}), F_1(s + \sqrt{A})],$$

which is to say

$$[F_1][F, s + \sqrt{A}] = [s + \sqrt{A}][F_1, G_1 + \sqrt{A}].$$

Therefore,

$$\begin{aligned}
&[F][s+\sqrt{A}][Y_1+X_1\sqrt{A}][B,g+\sqrt{A}] \\
&\qquad = [s+\sqrt{A}][F(Y_1+X_1\sqrt{A})][B,g+\sqrt{A}] \\
&\qquad = [s+\sqrt{A}][(s-\sqrt{A})(Y+X\sqrt{A})][B,g+\sqrt{A}] \\
&\qquad = [s^2-A][B][F,G+\sqrt{A}] \\
&\qquad = [F_1F][B][F,s+\sqrt{A}] \\
&\qquad = [F][s+\sqrt{A}][F_1,G_1+\sqrt{A}][B].
\end{aligned}$$

The very definition of equality of modules implies that the common factor $F(s+\sqrt{A})$ can be canceled from all terms in the lists at the beginning and end of this long equation to reach the desired conclusion $[Y_1+X_1\sqrt{A}][B,g+\sqrt{A}] = [B][F_1,G_1+\sqrt{A}]$.

Now let (X,Y) be a solution of $AX^2+B=Y^2$ in which X and Y are relatively prime. Then B and X must be relatively prime, so the congruence $Y+gX \equiv 0 \bmod B$ determines a square root g of $A \bmod B$ for which $[Y+X\sqrt{A}][B,g+\sqrt{A}] = [B(Y+X\sqrt{A}), Yg+AX+(Y+gX)\sqrt{A}] = [B][Y+X\sqrt{A}, R+S\sqrt{A}]$ where $R=\frac{Yg+AX}{B}$ and $S=\frac{Y+gX}{B}$. (Because $Yg+AX \equiv g(Y+gX) \equiv 0 \bmod B$, this expression for R defines a number.) Then $S(Y+X\sqrt{A}) = \frac{Y^2+gXY}{B} + SX\sqrt{A} = \frac{AX^2+B+gXY}{B} + SX\sqrt{A} = 1+X(R+S\sqrt{A})$, so $0 \equiv 1 \bmod [Y+X\sqrt{A}, R+S\sqrt{A}]$—that is, $[Y+X\sqrt{A}, R+S\sqrt{A}] = [1]$—and $[Y+X\sqrt{A}][B,g+\sqrt{A}] = [B]$ follows.

Let the reduction algorithm be applied to $[Y+X\sqrt{A}][B,g+\sqrt{A}] = [B]$ repeatedly to generate a sequence of equations $[Y_i+X_i\sqrt{A}][B,g+\sqrt{A}] = [B][F_i,G_i+\sqrt{A}]$, beginning with $F_0=1$ and ending, because $X=X_0>X_1>X_2>\cdots$, with an equation in which $X_N=0$, say $[Y_N][B,g+\sqrt{A}] = [B][F_N,G_N+\sqrt{A}]$. Since both sides are in canonical form, $Y_N=B=F_N$ and $g=G_N$.

Let the ith equation in this sequence be $[Y_i+X_i\sqrt{A}][B,g+\sqrt{A}] = [B][F_i,G_i+\sqrt{A}]$ so that $[F_0,G_0+\sqrt{A}]=[1]$ and $[F_N,G_N+\sqrt{A}] = [B,g+\sqrt{A}]$. As will be shown, *$[F_{i-1},G_{i-1}+\sqrt{A}]$ is the module that results from applying the comparison algorithm to $[F_i,G_i+\sqrt{A}]$ and the number used by the reduction algorithm* to go from the $(i-1)$st equation to the ith *is the number used by the comparison algorithm* to go from $[F_i, G_i + \sqrt{A}]$ to $[F_{i-1}, G_{i-1} + \sqrt{A}]$.

Let s_i be the number used by the reduction algorithm to go from the $(i-1)$st equation to the ith, and let t_i be the number used by the comparison algorithm to find the successor of $[F_i, G_i + \sqrt{A}]$. It is to be shown that the successor of $[F_i, G_i + \sqrt{A}]$ is $[F_{i-1}, G_{i-1} + \sqrt{A}]$ and that $s_i = t_i$ for $i = 1, 2, \ldots, N$.

Since $s_i^2 > A$ and $s_i + G_i \equiv 0 \bmod F_i$, the only way that s_i could fail to be t_i—the only way there could be a number smaller than s_i and congruent to s_i mod F_i whose square was greater than A—would be for s_i to be greater than F_i and for $(s_i - F_i)^2$ to be greater than A. If this were the case, it would follow that $s_i^2 + F_i^2 > 2s_iF_i + A$, $F_{i-1} + F_i > 2s_i$ (subtract A from both sides and divide by F_i), and $F_{i-1} > 2s_i - F_i > s_i$. Then $F_{i-1} - s_i$ would be a number greater than $s_i - F_i$ whose square was greater than A. In particular, $s_1 \neq t_1$ is impossible because $F_0 = 1$ implies that $F_0 > s_1$ is impossible.

If $i > 1$, $s_i \neq t_i$ would imply $F_{i-1} = s_i + t_{i-1}$ for the following reasons. Mod F_{i-1}, one has $s_i + t_{i-1} \equiv (s_i - G_{i-1}) + (t_{i-1} + G_{i-1}) \equiv 0 + 0$, so $s_i + t_{i-1}$ is divisible by F_{i-1}. If the quotient were not 1, it would follow that $s_i + t_{i-1} \geq 2F_{i-1}$. It was shown above that $F_{i-1} - s_i$ must be a number whose square is greater than A, so $t_{i-1} - F_{i-1} \geq F_{i-1} - s_i$ would also be a number whose square was greater than A, contrary to the definition of t_{i-1}. Therefore, $F_{i-1} = s_i + t_{i-1}$, as was to be shown.

Finally, if $i > 1$ and $s_{i-1} = t_{i-1}$, then $s_i = t_i$ because otherwise $F_{i-1} = s_i + t_{i-1} = s_i + s_{i-1}$. The formula

$$\frac{(s_{i-1} - \sqrt{A})(Y_{i-2} + X_{i-2}\sqrt{A})}{F_{i-2}} = Y_{i-1} + X_{i-1}\sqrt{A}$$

can be multiplied by $s_{i-1} + \sqrt{A}$ to find $F_{i-1}(Y_{i-2} + X_{i-2}\sqrt{A}) = (s_{i-1} + \sqrt{A})(Y_{i-1} + X_{i-1}\sqrt{A})$. In particular, $F_{i-1}X_{i-2} = s_{i-1}X_{i-1} + Y_{i-1}$. Since $X_{i-2} > X_{i-1}$, it would follow that $s_{i-1}X_{i-1} + Y_{i-1} > F_{i-1}X_{i-1} = s_iX_{i-1} + s_{i-1}X_{i-1}$ and $Y_{i-1} > s_iX_{i-1}$, contrary to the definition of s_i.

In summary, then, $s_i = t_i$ for $i = 1, 2, \ldots, N$ and application of the comparison algorithm N times to $[F_N, G_N + \sqrt{A}] = [B, g + \sqrt{A}]$ ends with $[F_0, G_0 + \sqrt{A}] = [1]$.

Multiply $(s_i - \sqrt{A})(Y_{i-1} + X_{i-1}\sqrt{A}) = F_{i-1}(Y_i + X_i\sqrt{A})$ by $s_i + \sqrt{A}$ and divide by F_{i-1} to find $F_i(Y_{i-1} + X_{i-1}\sqrt{A}) = (s_i + \sqrt{A})(Y_i + X_i\sqrt{A})$. The product of the N equations found in this way is

$$\prod_{i=0}^{N-1} (Y_i + X_i\sqrt{A}) \cdot F_N F_{N-1} \cdots F_1$$

$$= \prod_{i=1}^{N} (Y_i + X_i\sqrt{A}) \cdot (s_1 + \sqrt{A})(s_2 + \sqrt{A}) \cdots (s_N + \sqrt{A}).$$

Cancellation of the nonzero common factor $\prod_{i=1}^{N-1}(Y_i + X_i\sqrt{A})$ then gives $(Y + X\sqrt{A}) \cdot BF_{N-1}F_{N-2} \cdots F_1 = B \cdot (t_1 + \sqrt{A})(t_2 + \sqrt{A}) \cdots (t_N + \sqrt{A})$. In short, the solution $Y + X\sqrt{A} = \frac{(t_1+\sqrt{A})(t_2+\sqrt{A})\cdots(t_N+\sqrt{A})}{F_1F_2\cdots F_{N-1}}$ is the one that results when the method of the theorem of Chapter 19 is applied to $[B, g + \sqrt{A}]$ and N steps of the comparison algorithm are taken.

□

Exercises for Chapter 20

1. For each of the solutions of $A\square + B = \square$ that you found in the exercises of Chapter 19, use the reduction algorithm to retrace the steps from $[B, g + \sqrt{A}]$ to $[1]$.

2. Prove the variation of Brahmagupta's formula alluded to in the proof of Theorem 3.

3. Find the contents of several of the stable modules listed in the appendix (page 169), focusing, of course, on the modules that are not primitive.

Exercises on Fundamental Units

The exercises that follow draw on some simple concepts of algebraic number theory. This is not the place to enter into explanations of them, but readers already familiar with them, or readers willing to study them in books on algebraic number theory, may find the exercises interesting and challenging. Moreover, only a very little understanding of these concepts is necessary to understand the calculations, which in the end merely involve solutions of $A\Box + B = \Box$. Exercise 13 below gives an interesting simplification of the computation of the solution of Pell's equation $A\Box + 1 = \Box$ in the most difficult cases.

A **quadratic number field** is a field that is obtained by adjoining to the field of rational numbers one root of one irreducible quadratic polynomial with rational coefficients. Elementary considerations (complete the square) show that every such field can be put in the form $\mathbf{Q}(\sqrt{A})$ where A is an integer (possibly negative) that is not 0 or 1 and is divisible by no square greater than 1. Operationally, such a field is the set of expressions of the form $y + x\sqrt{A}$ in which x and y are *rational numbers*. They are added, subtracted, multiplied, and divided in the usual ways.

An element of a quadratic number field is called an **algebraic integer** if it is a root of a polynomial of the form $X^2 + aX + b$ in which a and b are ordinary integers (that is, either a or $-a$ is a number in the sense of Chapter 1 and the same is true of b). A nonzero element of a quadratic number field is called a **unit** if both it and its reciprocal are algebraic integers.

The fundamental theorem about units of quadratic number fields states that *each quadratic number field contains a unit ε with the property that every unit in the field can be written in the form $\pm\varepsilon^n$ for some choice of the sign in front and for some integer n.* (Again, to say that n is an integer means that either n or $-n$ is a number.) Such an ε is called a **fundamental unit** of the quadratic number field to which it belongs.

The following exercises are concerned with the construction of a fundamental unit (and therefore of *all* units) of each quadratic number field $\mathbf{Q}(\sqrt{A})$, where A is divisible by no square greater than 1 and $A \neq 0$ or 1.

4. Show that $y + x\sqrt{A}$ is an algebraic integer when x and y are ordinary integers. A great step forward in the history of algebraic number theory occurred in the middle of the 19th century when the importance of the notion of algebraic integers became clear, along with the realization that there are some algebraic integers $y + x\sqrt{A}$ in which x and y are *not* integers. Prove that $\frac{1+\sqrt{A}}{2}$ is an algebraic integer when $A \equiv 1 \bmod 4$.

A basic theorem states that sums and products of algebraic integers are algebraic integers. Therefore, Exercise 4 implies that $\frac{y+x\sqrt{A}}{2}$ is an algebraic integer whenever $A \equiv 1 \bmod 4$ and x and y are both odd. Another basic theorem of algebraic number theory implies that *the cases listed give a complete catalog of the algebraic integers in the field* $\mathbf{Q}(\sqrt{A})$; that is, the algebraic integers in $\mathbf{Q}(\sqrt{A})$ are simply the elements $y + x\sqrt{A}$ in which x and y are ordinary integers, except that when $A \equiv 1 \bmod 4$, the elements in which both x and y are odd integers divided by 2 are also algebraic integers. The problem is to determine which of these integers are units and to show that one of them is a fundamental unit.

5. Prove that if an algebraic integer is a root of a polynomial of the form $X^2 + cX + d$ in which c and d are rational numbers, then c and d must be integers.

6. Prove that if $y+x\sqrt{A}$ is a unit of $\mathbf{Q}(\sqrt{A})$, then $y^2 - Ax^2 = \pm 1$. [The rational number $y^2 - Ax^2$, called the **norm** of $y + x\sqrt{A}$, is the coefficient b of the polynomial $X^2 + aX + b$ of which $y + x\sqrt{A}$ is a root.]

7. Find all units in all fields $\mathbf{Q}(\sqrt{A})$ in which A is negative.

For positive A, each unit of $\mathbf{Q}(\sqrt{A})$ determines a solution of $Ax^2 + 1 = y^2$ or $Ax^2 = y^2 + 1$ or, when $A \equiv 1 \bmod 4$, perhaps a solution of $Ax^2 + 4 = y^2$ or $Ax^2 = y^2 + 4$ in which x and y are *numbers*; conversely, the numbers that are determined in this way determine the coefficients of the corresponding unit, except for their signs, so the theorem of Chapter 19 suffices to find all units.

8. A solution of $Ax^2 = y^2 + 1$ implies a solution of $z^2 = Ay^2 + A$ in which $z \equiv 0 \bmod A$. The theorem of Chapter 19 tells how to find all solutions of $Ay^2 + A = z^2$. Show that if this equation has

a solution then repeated application of the comparison algorithm to [1] goes through an even number of steps before returning to [1]; in fact, the first half of the steps go from [1] to $[A, \sqrt{A}]$ and the second half of the steps return in a symmetrical way from $[A, \sqrt{A}]$ to [1]. Moreover, the smallest solution of Pell's equation for such an A is given by a hypernumber of the form $y_1 + x_1\sqrt{A} = \frac{(z_2+y_2\sqrt{A})^2}{A}$ where $z_2 + y_2\sqrt{A}$ gives the smallest solution of $z_2^2 = Ay_2^2 + A$. Prove that $z_2 \equiv 0 \bmod A$, so that the smallest solution of Pell's equation can in fact be written in the form $(y_2+x_2\sqrt{A})^2$ where $x_2 = z_2/A$ and where, consequently, the coefficients of $y_2 + x_2\sqrt{A}$ give the smallest solution of $Ax_2^2 = y_2^2 + 1$.

9. Prove that if $Ax^2 = y^2 + 1$ has a solution (in numbers), then the hypernumber $\varepsilon = y_2 + x_2\sqrt{A}$, where $Ax_2^2 = y_2^2 + 1$ is the *smallest* solution, has the property that all units whose coefficients are integers have the form $\pm\varepsilon^n$ for integer n, and that otherwise the hypernumber $\varepsilon = y_1 + x_1\sqrt{A}$, where $Ax_1^2 + 1 = y_1^2$ is the smallest solution of Pell's equation, has that property.

10. It remains to find the units that have a 2 in the denominator. The condition $A \equiv 1 \bmod 4$ is necessary for $\mathbf{Q}(\sqrt{A})$ to contain such units. Prove the stronger necessary condition $A \equiv 5 \bmod 8$.

11. Show that when $A = 37$, there are no units of the form $\frac{y+x\sqrt{A}}{2}$. More generally, how can one determine, for a given number $A \equiv 5 \bmod 8$, whether $\mathbf{Q}(\sqrt{A})$ contains such units?

12. If $[4, 1+\sqrt{A}]$ is in the cycle of [1] and and $[A, \sqrt{A}]$ is not in the cycle of [1], the *cube* of a fundamental unit gives a unit $y_1+x_1\sqrt{A}$ that describes the smallest solution of Pell's equation. Give an algorithm for constructing such a fundamental unit and apply it in the cases $A = 21$ and 69.

13. Finally, when $A \equiv 5 \bmod 8$ and both $[A, \sqrt{A}]$ and $[4, 1+\sqrt{A}]$ are in the cycle of [1], the hypernumber $y_1 + x_1\sqrt{A}$ whose coefficients give the smallest solution of Pell's equation is the *sixth* power of a fundamental unit. A method for finding a fundamental unit that seems to be effective in these cases is to note that the modules $[A, \sqrt{A}][4, 1 + \sqrt{A}]$ and $[A, \sqrt{A}][4, 3 + \sqrt{A}]$ must both be in the cycle of [1]. When the method of Chapter 19 for writing a module in the

cycle of [1] in the form $[y + x\sqrt{A}]$ (where $y^2 > Ax^2$) is used, one of these two modules turns out to be much easier to write in this form than the other. Then division by $\sqrt{4A} = 2\sqrt{A}$ gives a fundamental unit. Use this method to find a fundamental unit in the first seven cases of this type, namely, $A = 5$, 13, 29, 53, 61, 85, and 109. (The value $A = 45$ is skipped because $\mathbf{Q}(\sqrt{45}) = \mathbf{Q}(\sqrt{5})$.) Note that this gives a much shorter computation of the solution of Pell's equation for $A = 61$ than Exercise 4 of Chapter 19 does.

Chapter 21

Euler's Remarkable Discovery

The solution of $A\square + B = \square$ begins by finding all possible square roots of A mod B. In particular, it requires determining whether A *has* a square root mod B. Just as testing a number for primality is a very different and much easier problem than factoring the number, so is determining whether a number A has a square root mod B a different and easier problem than finding a square root of A mod B.

When B is a product $B = B_1B_2$, a square root of A mod B_1B_2 determines a square root of A mod B_1 and a square root of A mod B_2. When B_1 and B_2 are relatively prime, the Chinese remainder theorem implies that, conversely, a square root of A mod B_1 and a square root of A mod B_2 can be combined to find a square root of A mod B_1B_2. Therefore, the problem of finding square roots of A mod B reduces to the case in which B is a power of a prime number. Since a square root of A mod p^e for $e > 1$ is a square root of A mod p, the solution of $A\square + B = \square$ leads in this way to:

Problem. *Given a number A, not a square, for which primes p is A a square mod p?*

For a given p, it is easy to determine the numbers A that are squares mod p in the following way. Given a prime $p > 2$ and a number A, not a square, let $C_p(A)$ denote what is called the **quadratic**

character of $A \bmod p$, which is 1, 0, or -1 according to the rule: $C_p(A) = 1$ if A is congruent to a nonzero square mod p, $C_p(A) = 0$ if A is zero mod p, and $C_p(A) = -1$ otherwise.[1]

Proposition. $C_p(A) \equiv A^{(p-1)/2} \bmod p$.

This statement is often called **Euler's criterion** for determining whether A is a square mod p.

Proof. Let g be a primitive root mod p. If $A \equiv 0 \bmod p$, then of course $A^{(p-1)/2} \equiv 0 \equiv C_p(A) \bmod p$. Otherwise, $A \equiv g^i \bmod p$ where i is the index of A with respect to g. Since A is a square mod p if and only if[2] i is even, $A^{(p-1)/2} \equiv g^{i(p-1)/2} \equiv (-1)^i = C_p(A) \bmod p$. □

Corollary. *For any prime $p > 2$ and for any numbers A and B,*

$$C_p(AB) = C_p(A)C_p(B).$$

Deduction. From $(AB)^{(p-1)/2} = A^{(p-1)/2}B^{(p-1)/2}$ it follows that $(AB)^{(p-1)/2} \equiv A^{(p-1)/2}B^{(p-1)/2} \bmod p$, so the proposition implies $C_p(AB) \equiv C_p(A)C_p(B) \bmod p$, from which $C_p(AB) = C_p(A)C_p(B)$ follows. □

But $C_p(A)$ tells which numbers A are squares modulo a fixed p and therefore does not address the problem above, in which A is fixed and p varies over all possible primes. Euler studied the problem around 1753 and discovered *empirically* the amazing fact that

(1) the value of $C_p(A)$ depends only on the value of $p \bmod 4A$.

In other words, if p and q are primes and if $p \equiv q \bmod 4A$, then $C_p(A) = C_q(A)$. Euler was able to test enough cases to be thoroughly convinced that this rule held in all cases, even though he realized that it is an open-ended statement that could never be *proved* empirically.

[1]Strictly speaking, then, $C_p(A)$ is not a number, as far as this book is concerned, because -1 is not a number. One can smooth this over by thinking of the values of $C_p(A)$ as being 0, 1, or 3 mod 4. All that matters is the multiplication table: $0 \cdot 0 = 0$, $0 \cdot 1 = 0$, $0 \cdot (-1) = 0$, $1 \cdot 1 = 1$, $1 \cdot (-1) = -1$, and $(-1) \cdot (-1) = 1$, which is the multiplication table of 0, 1, and 3 mod 4.

[2]See the proposition of Chapter 15.

(Note that if p divides $4A$, then $p = 2$ or p divides A, so $p \equiv q \bmod 4A$ holds only when $q = p$, which shows that (1) is trivially true for such primes.)

Actually, Euler's reasons for being convinced of the truth of (1) went beyond simple testing of individual values of $C_p(A)$ for fixed A and varying p. He went on to develop more detailed conjectures about the values of $C_p(A)$, the chief one of which is that

if p, q, and r are primes for which $pq \equiv r \bmod 4A$ then

$$C_p(A)C_q(A) = C_r(A). \tag{2}$$

This statement superficially resembles the corollary above, but it is a far different and more surprising statement; note, for example, that pq is not prime, so (2) is a statement about primes r congruent to $pq \bmod 4A$ when there is in fact no obvious reason to believe that there even *is* a prime r congruent to $pq \bmod 4A$. A third observation comes to light very quickly when (1) and (2) are explored, namely,

(3) $C_p(A)$ takes the value 1 for exactly half of the possible values of $p \bmod 4A$ and the value -1 for the other half.

(when one excludes the few primes that divide $4A$ and when one assumes, as is assumed throughout, that A is not a square). That is, one can partition the $\phi(4A)$ numbers less than $4A$ and relatively prime to $4A$ into two sets of equal size in such a way that $C_p(A) = 1$ for primes p congruent to numbers in one set and $C_p(A) = -1$ for primes p congruent to numbers in the other.[3]

When one explores the implications of these three conjectures—see the exercises—one develops a tightly woven web of methods for predicting the values of $C_p(A)$ for a given A that is always borne out by computation. It is in this way that Euler could convince himself

[3]In the language of group theory, the numbers less than $4A$ and relatively prime to $4A$ form a *group* of order $\phi(4A)$. Statements (1), (2), and (3) say that there is a subgroup of index 2 in this group with the property that $C_p(A) = 1$ if and only if the class of $p \bmod 4A$ is in this subgroup. There is no assertion here that every number relatively prime to $4A$ is congruent to a prime mod $4A$ but a famous theorem of Dirichlet proves that this is in fact true.

and his readers that the statements were true, even though he could not prove them.

A phenomenon of this sort—a detailed, simple prediction that holds up under extensive testing—suggests that there will be a simple explanation that can be discovered by looking at the phenomenon in a different way. No doubt Euler in 1753 looked for a simple explanation. But he didn't find one. He lived until 1783, and some of his investigations focused very narrowly on questions of exactly this type, but he was never able to prove his conjectures of 1753.

The young Gauss published not one but two proofs of them in the *Disquisitiones Arithmeticae* in 1801, but although he succeeded after considerable effort in proving that the statements were correct, he does not seem to have *explained* them to his satisfaction, because he continued to study questions related to them as he searched for a deeper understanding of the phenomenon. His search resulted in many intriguing observations and alternative statements, but a certain mystery remained and still remains about the "real explanation."

Euler's conjectures acquired the imposing name of The Law of Quadratic Reciprocity during the 19th century and came to be seen as a major focus of elementary (and not-so-elementary) number theory. The reason for the term "reciprocity law" will be seen in Chapter 27. But in more modern times the "reciprocity" feature has come to seem less important, and modern "generalized reciprocity laws" do not in fact take the form of a statement about reciprocity. Instead, Euler's original observation (1) seems to be the core idea. In order to have a convenient name for it, in the chapters that follow it will be called:

Euler's Law. *Let p, q, and A be numbers, A not a square. If p and q are prime and if $p \equiv q \bmod 4A$, then $C_p(A) = C_q(A)$.*

Exercises for Chapter 21

Study Questions.

1. For $A = 2$, 3, 5, and 6, list the $\phi(4A)$ numbers that are less than $4A$. For each listed number, find at least two primes that are congruent to it mod $4A$ and use Euler's criterion to determine whether

A is a square modulo them. Verify that $C_p(A)$ has the same value for all the primes p you have listed that are congruent to the same number mod $4A$. If the common value is 1, draw a circle around the number and otherwise draw a square around it. Note that Euler's properties (2) and (3) are valid for these four values of A. There is a *fourth* fact about the distribution of circles and squares that relates the values of $C_p(A)$ and $C_q(p)$ when $p+q \equiv 0 \bmod 4A$. Find it. Carry the process beyond $A = 6$.

2. For larger values of A, try finding the numbers that get circles using as few tests as possible. By Euler's property (2), any number less than $4A$ and relatively prime to $4A$ that is a *square* mod $4A$ must get a circle. Thus, 1, 9, 25, 49, ... all get circled as long as they are less than $4A$. So do $4A - 1$, $4A - 9$, ... For $A = 2$, 3, and 5, these simple observations account for *all* of the circled numbers, because these observations imply that half of the numbers must be circled, so the remaining numbers must all get squares. For $A = 6$ one needs only one test—namely, the easy observation that 6 is a square mod 5—to find that 5 should be circled, after which all circled numbers are accounted for. How many tests do you need when $A = 15$? Try others.

3. Show that if A is prime, one can determine the circled numbers by the method of Exercise 2 without doing *any* tests. The resulting statement is one version of the law of quadratic reciprocity.

Computations.

4. Choose a 5-digit number A, not a square, and find a 10-digit prime p. Find a number j for which $p + 4jA$ is prime, call it q, and then apply Euler's criterion to verify that $C_p(A) = C_q(A)$.

Chapter 22

Stable Modules

The comparison algorithm applied to a module $[f, g+\sqrt{A}]$ in canonical form (with $e = 1$) determines another module $[f_1, g_1+\sqrt{A}]$ in the same form, the **successor** of $[f, g + \sqrt{A}]$. The **sequence of successors** of $[f, g + \sqrt{A}]$ is the infinite sequence $[f_1, g_1 + \sqrt{A}]$, $[f_2, g_2 + \sqrt{A}]$, ... in which each new module is the successor of the one that came before. A module $[f, g + \sqrt{A}]$ will be called **stable** if it occurs in its own sequence of successors, so that the sequence of successors cycles back to $[f, g + \sqrt{A}]$ itself and then endlessly repeats the same cycle.

Theorem. *Let A be a given number, not a square. For each number k whose square is less than A and for each factor f of $A - k^2$ that satisfies $f \geq 2k$, the modules $[f, k + \sqrt{A}]$ and $[f, f - k + \sqrt{A}]$ are stable modules, and every stable module for this A has this form. In particular, the number of stable modules is twice the number of pairs (k, f) that satisfy these requirements (that $k^2 < A$, $k^2 \equiv A \bmod f$, and $f \geq 2k$) minus the number of such pairs for which $[f, k + \sqrt{A}] = [f, f - k + \sqrt{A}]$ (or, more specifically, minus the number of such pairs in which*[1] *$k = 0$ or $2k = f$).*

[1]When $k = 0$, $[f, k+\sqrt{A}] = [f, f - k+\sqrt{A}]$ but the latter is not in canonical form. When $k > 0$ and $f = 2k$, $[f, k + \sqrt{A}]$ and $[f, f - k + \sqrt{A}]$ are identical. Otherwise, $[f, k+\sqrt{A}]$ and $[f, f-k+\sqrt{A}]$ are in canonical form and not identical, so they are distinct.

Proof. Let $\mathcal{M}$ denote the set of all modules $[f, g+\sqrt{A}]$ for which $|r-f|^2 < A$, where r is the number used by the comparison algorithm to find the successor of $[f, g+\sqrt{A}]$—that is, where r is the smallest solution of $r+g \equiv 0 \bmod f$ for which $r^2 > A$. It was shown in Chapter 20 that $\mathcal{M}$ is a finite set and that the successor of a module in $\mathcal{M}$ is in $\mathcal{M}$. Thus, the function that assigns to each module in $\mathcal{M}$ its successor is a function from the finite set $\mathcal{M}$ to itself.

This function is *one-to-one* because the successor $[f_1, g_1+\sqrt{A}]$ of $[f, g+\sqrt{A}]$ determines $[f, g+\sqrt{A}]$ in the following way. First, r is the least number congruent to g_1 mod f_1 for which $r^2 > A$, because if it were not, then r would be greater than f_1 and $(r-f_1)^2$ would be greater than A, which would imply $r^2+f_1^2 > 2rf_1+A$, $ff_1+f_1^2 > 2rf_1$, $f+f_1 > 2r$, $f^2+ff_1 > 2rf$, $f^2+r^2 > 2rf+A$, contrary to the assumption $|r-f|^2 < A$. Then f is determined by $f = (r^2-A)/f_1$ and g is determined as the least solution of $g+r \equiv 0 \bmod f$.

Thus, the successor function, being a one-to-one function from a finite set to itself, is an *onto* function, which is to say that it is simply a permutation of $\mathcal{M}$. It therefore partitions $\mathcal{M}$ into *cycles,* and the sequence of successors of any module in $\mathcal{M}$ is a cycle, so all modules in $\mathcal{M}$ are stable.

No other modules are stable, because, as was shown in Chapter 20, the value of $|r-f|^2$ for a module not in $\mathcal{M}$ decreases for successive modules until $\mathcal{M}$ is reached. What is to be shown, then, is that the description in the theorem describes the modules of $\mathcal{M}$.

The modules $[f, k+\sqrt{A}]$ or $[f, f-k+\sqrt{A}]$ described in the theorem are in $\mathcal{M}$. In the case of $[f, k+\sqrt{A}]$, $r = jf-k$ for some $j \geq 1$; if $j=1$, then $k = f-r$, so $|r-f|^2 = k^2 < A$, but if $j > 1$, then $r \geq 2f-k = f+(f-k) > f$ and $(r-f)^2 < A$ by the definition of r. In the case of $[f, f-k+\sqrt{A}]$, r has the form $r = jf+k$ for some j; the condition $k^2 < A$ implies $j \neq 0$, so $r \geq f$ and $(r-f)^2 < A$ by the definition of r.

Finally, a module in $\mathcal{M}$ is one of those described by the theorem, as can be seen in the following way. To say that $[f, g+\sqrt{A}]$ is in $\mathcal{M}$ means that $|r-f|^2 < A$, where r is defined as before. Let $l = |r-f|$. Then $l^2 < A$ and $l^2 = r^2-2rf+f^2 \equiv r^2 \equiv A \bmod f$. If $l \geq f$, then

$l_1 = l - f$ satisfies $l_1^2 < l^2 < A$ and $l_1^2 \equiv l^2 \equiv A \bmod f$. Similarly, if $l_1 \geq f$, then $l_2 = l_1 - f$ satisfies $l_2^2 < A$ and $l_2^2 \equiv A \bmod f$. Continuing in this way, one must eventually reach an l_i for which $l_i^2 < A$, $l_i^2 \equiv A \bmod f$, and $l_i < f$. Let $m = l_i$. If $2m \leq f$, then $[f, m + \sqrt{A}]$ and $[f, f - m + \sqrt{A}]$ are both among the modules listed by the theorem. If $2m > f$, then $m' = f - m$ satisfies $2m' = 2f - 2m = f - (2m - f) < f$, as well as $(m')^2 = f^2 - 2mf + m^2 < f^2 - f^2 + m^2 = m^2 < A$ and $(m')^2 \equiv m^2 \equiv A \bmod f$, so both $[f, m' + \sqrt{A}] = [f, f - m + \sqrt{A}]$ and $[f, f - m' + \sqrt{A}] = [f, m + \sqrt{A}]$ are listed by the theorem. The given module is $[f, r + \sqrt{A}]$ (not necessarily in canonical form); since $l = |r - f| \equiv \pm r \bmod f$ and $m \equiv l \bmod f$, the given module is either $[f, m + \sqrt{A}]$ or $[f, f - m + \sqrt{A}]$, both of which are listed by the theorem. □

Once the stable modules have been listed, the comparison algorithm can be used to partition them into cycles.

Exercise for Chapter 22

The table in the appendix (page 169) gives one stable module from each cycle of stable modules for each value of A, not a square, less than 112. Choose a value of A, use the theorem of this chapter to find all stable modules for the chosen A, and use the comparison algorithm to find how they are partitioned into cycles. Then compare your answer to the table in the appendix. (The meaning of the + and − signs in the tables will be explained in Chapter 24.)

Chapter 23

Equivalence of Modules

As before, let A be a fixed number, not a square, and let all hypernumbers and all modules of hypernumbers be understood to derive from this A.

Definitions. *A module is* **principal** *if it can be expressed in the form* $[y+x\sqrt{A}]$ *where* x *and* y *are numbers for which* $Ax^2 < y^2$*. Two modules* M_1 *and* M_2 *are* **equivalent**, *written* $M_1 \sim M_2$, *if there are principal modules* P_1 *and* P_2 *for which* $P_1M_1 = P_2M_2$.

Brahmagupta's formula implies that *a product of principal modules is principal* because the product of $[y + x\sqrt{A}]$ and $[v + u\sqrt{A}]$ is $[(yv + Axu) + (yu + xv)\sqrt{A}]$ and Brahmagupta's formula states that $A(yu + xv)^2 + (y^2 - Ax^2)(v^2 - Au^2) = (yv + Axu)^2$, so $y^2 > Ax^2$ and $v^2 > Au^2$ imply $(yv + Axu)^2 > A(xv + yu)^2$. This property of principal modules is used in the proof that "equivalence," as it is defined above, is transitive: If $M_1 \sim M_2$ and $M_1 \sim M_3$, there are principal modules P_1, P_2, Q_1, and Q_2 for which $P_1M_1 = P_2M_2$ and $Q_1M_1 = Q_2M_3$; then $P_2Q_1M_2 = P_1Q_1M_1 = P_1Q_2M_3$ and P_2Q_1 and P_1Q_2 are principal, which shows that $M_2 \sim M_3$.

As is obvious from the definition, "equivalence" is also reflexive and symmetric, so it is a true equivalence relation. Moreover, it is consistent with the multiplication of modules in the sense that $M_1 \sim M_2$ implies $M_1M_3 \sim M_2M_3$ for all modules M_3.

Problem. *Given two modules, determine whether they are equivalent.*

The comparison algorithm produces an equation $[r+\sqrt{A}][f, g+\sqrt{A}] = [f][f_1, g_1+\sqrt{A}]$ which shows that $[f_1, g_1+\sqrt{A}] \sim [f, g+\sqrt{A}]$ because $[r+\sqrt{A}]$ and $[f]$ are principal modules. (Recall that $r^2 > A$ by the choice of r.) Therefore, a module is equivalent to all of its successors. In particular, *every module is equivalent to a stable module.* Therefore, to be able to determine whether two modules are equivalent, it will suffice to be able to determine whether two *stable* modules are equivalent.

Since a stable module is equivalent to its successors, a *sufficient* condition for two stable modules to be equivalent is for them to be in the same cycle. The problem will be solved by proving that this sufficient condition is also necessary:

Theorem. *Equivalent stable modules lie in the same cycle.*

A module $[e][f, g+\sqrt{A}]$ in canonical form is equivalent to $[f, g+\sqrt{A}]$. Therefore, the assumption that two modules in canonical form $[e][f, g+\sqrt{A}]$ and $[E][F, G+\sqrt{A}]$ are equivalent is the assumption that there are hypernumbers $s+r\sqrt{A}$ and $v+u\sqrt{A}$ for which $s^2 > Ar^2$ and $v^2 > Au^2$ and $[s+r\sqrt{A}][f, g+\sqrt{A}] = [v+u\sqrt{A}][F, G+\sqrt{A}]$. This assumption can be put in simpler form using:

Lemma. *Given a principal module $[v+u\sqrt{A}]$, there is a principal module $[V+U\sqrt{A}]$ for which $[V+U\sqrt{A}][v+u\sqrt{A}] = [n]$ for some nonzero number n.*

Proof. Since $[1]$ is a stable module, it occurs infinitely often in the sequence of its successors and the theorem of Chapter 19 implies that the equation $Ax^2+1 = y^2$ has infinitely many solutions. Moreover, the numbers x and y in successive solutions grow without bound, so there is a solution of this equation in which $x > u$. Then $(uy)^2 = u^2(Ax^2+1) < Au^2x^2 + x^2 = (Au^2+1)x^2 \leq (vx)^2$ (because $Au^2 < v^2$ implies $Au^2+1 \leq v^2$). In short, $uy < vx$. Furthermore, $v^2y^2 > (Au^2)(Ax^2) = (Aux)^2$, so $vy > Aux$. Therefore, $(y+x\sqrt{A})(v-u\sqrt{A})$ is a well-defined hypernumber in the sense that $(y+x\sqrt{A})u\sqrt{A}$ can be subtracted from $(y+x\sqrt{A})v$, say $V+U\sqrt{A} = (y+x\sqrt{A})(v-u\sqrt{A})$. Then $[V+U\sqrt{A}][v+u\sqrt{A}] = [(y+x\sqrt{A})(v-u\sqrt{A})(v+u\sqrt{A})] = [(y+x\sqrt{A})(v^2-Au^2)] = [v^2-Au^2][y+x\sqrt{A}] = [v^2-Au^2]$ (because $1 = y^2-Ax^2$ implies $[y+x\sqrt{A}] = [1, y+x\sqrt{A}] = [1]$), which completes the proof of the lemma. □

Proof of the theorem. The lemma shows that it will suffice to prove that if $[f, g+\sqrt{A}]$ and $[F, G+\sqrt{A}]$ are stable modules in canonical form and if there is a hypernumber $Y+X\sqrt{A}$ for which $Y^2 > AX^2$ and $[Y+X\sqrt{A}][f, g+\sqrt{A}] = [E][F, G+\sqrt{A}]$, then $[f, g+\sqrt{A}]$ and $[F, G+\sqrt{A}]$ are in the same cycle.

Because the equation can be multiplied by $[f]$, if needed, one can assume without loss of generality that E is divisible by f—that is, that the given equation has the form $[Y+X\sqrt{A}][f, g+\sqrt{A}] = [fE][F, G+\sqrt{A}]$. Then $(Y+X\sqrt{A})f \equiv 0 \bmod [fE][F, G+\sqrt{A}]$ implies $fX \equiv 0 \bmod fE$ and $fY \equiv GfX \equiv 0 \bmod fEF$ by the corollary of Chapter 18. Since these congruences imply $X \equiv 0 \bmod E$ and $Y \equiv GX \bmod EF$, they imply that both X and Y are zero mod E and division by $[E]$ puts the given equation in the form

$$[Y+X\sqrt{A}][f, g+\sqrt{A}] = [f][F, G+\sqrt{A}] \tag{1}$$

for a new hypernumber $Y+X\sqrt{A}$ in which $Y^2 > AX^2$. In short, the equivalence of $[f, g+\sqrt{A}]$ and $[F, G+\sqrt{A}]$ implies an equivalence of the special form (1).

If $X = 0$, equation (1) implies that $[f, g+\sqrt{A}]$ and $[F, G+\sqrt{A}]$ are identical. Otherwise, let the reduction algorithm be applied to (1) as in Chapter 20 to find a sequence of equations $[Y_i+X_i\sqrt{A}][f, g+\sqrt{A}] = [f][F_i, G_i+\sqrt{A}]$ in which $X = X_0 > X_1 > \cdots > X_N = 0$. As before, let s_i be the number used by the reduction algorithm to go from the $(i-1)$st equation to the ith and let t_i be the number used by the comparison algorithm to find the successor of $[F_i, G_i+\sqrt{A}]$. By the same argument as before, $s_i \neq t_i$ would imply $F_{i-1} > s_i$ and $(F_{i-1}-s_i)^2 > A$. This is impossible when $i = 1$ because then $x = F_0 - s_1$ would be the smallest solution of $x + G_0 \equiv 0 \bmod F_0$ and $(F_0 - s_1)^2 > A$ would contradict the assumption that $[F_0, G_0+\sqrt{A}]$ is stable. Also as before, $s_{i-1} = t_{i-1}$ for $i > 1$ implies $s_i = t_i$. Therefore,

$[F, G+\sqrt{A}] = [F_0, G_0+\sqrt{A}]$ is the Nth successor of $[F_N, G_N+\sqrt{A}] = [f, g+\sqrt{A}]$. In particular, $[F, G+\sqrt{A}]$ is in the cycle of $[f, g+\sqrt{A}]$, as was to be shown. □

Thus, equivalence classes of modules for a given A correspond one-to-one to cycles of stable modules for that A. The principal modules are those that are equivalent to $[1]$. (In fact, they are called "principal" because they are in the *principal equivalence class,* the class of the module $[1]$.) Because there is a natural way to multiply equivalence classes of modules—the equivalence class of a product of two modules is determined by the equivalence classes of the factors—it follows that there is a natural way to multiply cycles. This multiplication of cycles is commutative and associative, and the **principal cycle**, the cycle of $[1]$, is an identity with respect to this multiplication.

Recall that a module is **primitive** if its content is 1 (Chapter 20).

Proposition. *A module is primitive if and only if its equivalence class is* invertible *relative to the multiplication of equivalence classes that was just defined. Otherwise stated, a given module is primitive if and only if there is a module whose product with the given module is equivalent to* $[1]$.

Proof. A primitive module $[e][f, g+\sqrt{A}]$ is invertible because its norm—it product with its conjugate—is $[e^2 f]$, which is equivalent to $[1]$. That the converse holds—an invertible module is primitive—can be proved in the following way:

Let $[e^2 f][d, g+\sqrt{A}]$ be the norm of $[e][f, g+\sqrt{A}]$ as in Chapter 20. Then $[d, g+\sqrt{A}][f, g+\sqrt{A}] = [df, d(g+\sqrt{A}), f(g+\sqrt{A}), (g+\sqrt{A})^2] = [df, d(g+\sqrt{A}), g^2+A+2g\sqrt{A}]$ because f is a multiple of d. Moreover, $2g$ is a multiple of d, say $2g = qd$, and subtraction of q times the middle term from the last gives $g^2+A-qdg = g^2+A-2g^2 \equiv 0 \bmod f$, so the last term can be dropped, leaving $[d, g+\sqrt{A}][f, g+\sqrt{A}] = [df, d(g+\sqrt{A})] = [d][f, g+\sqrt{A}]$. Therefore, $[d, g+\sqrt{A}][f, g+\sqrt{A}] \sim [f, g+\sqrt{A}]$. If $[e][f, g+\sqrt{A}]$ is invertible, there is a module $[E][F, G+\sqrt{A}]$ whose product with $[e][f, g+\sqrt{A}]$ is equivalent to [1], so multiplication of $[d, g + \sqrt{A}][f, g + \sqrt{A}] \sim [f, g + \sqrt{A}]$ by $[e][E][F, G + \sqrt{A}]$ then gives $[d, g + \sqrt{A}] \sim [1]$.

Thus, if $[e][f, g + \sqrt{A}]$ is invertible, the theorem of this chapter implies that [1] is a successor of $[d, g + \sqrt{A}]$. Since it was shown in Chapter 20 that no successor of a module that is not primitive can be primitive, the desired conclusion $d = 1$ follows. □

For readers familiar with the terminology of group theory: The primitive cycles of stable modules form a finite commutative *group* under multiplication; this group is called the **class group** for the given A.

Exercise for Chapter 23

1. Prove that the equivalence class of $[3, \sqrt{75}]$ is invertible by (a) finding its content, (b) finding its square, and (c) applying the comparison algorithm to its product with its successor.

Chapter 24

Signatures of Equivalence Classes

In studying "Euler's law" (see Chapter 21), it is natural to assume that A is not divisible by any square, because $C_p(n^2A) = C_p(A)$ except for the few primes p that divide n, which shows that Euler's law for A implies it for n^2A. (If $p \equiv q \bmod 4n^2A$, then $p \equiv q \bmod 4A$, which implies $C_p(A) = C_q(A)$ and therefore implies $C_p(n^2A) = C_q(n^2A)$ unless $n \equiv 0 \bmod p$, in which case $C_p(n^2A) = C_q(n^2A) = 0$.) In other words, it is natural to assume that A is a product of distinct primes. Such an A is called **squarefree**.

Proposition 1. *Given an odd prime factor A_1 of a squarefree number A and given a module $[f, g + \sqrt{A}]$ for that A, there is a module $[f', g' + \sqrt{A}]$ in canonical form equivalent to $[f, g + \sqrt{A}]$ in which $f' \not\equiv 0 \bmod A_1$. The value of $C_{A_1}(f')$ is the same for all such modules $[f', g' + \sqrt{A}]$.*

Proof. Assume without loss of generality that $[f, g+\sqrt{A}]$ is in canonical form and let $[f_1, g_1 + \sqrt{A}]$ be its successor. Then $ff_1 = r^2 - A$ for the r used by the comparison algorithm to determine the successor. Suppose A_1 divides f. Since A_1 divides A, A_1 must then divide r^2, and since A_1 is prime, it must divide r and therefore A_1^2 must divide r^2; if A_1 divided f_1, then A_1^2 would divide $A = r^2 - ff_1$, which is impossible because A is squarefree. In short, if $[f, g + \sqrt{A}]$ fails to have the property that $f \not\equiv 0 \bmod A_1$, then its successor does have that property, which proves the first statement of the proposition.

If neither f nor f_1 is divisible by A_1, then $C_{A_1}(f) = C_{A_1}(f_1)$ because $ff_1 = r^2 - A \equiv r^2 \bmod A_1$; therefore $C_{A_1}(f)C_{A_1}(f_1) = C_{A_1}(ff_1) = C_{A_1}(r^2 - A) = C_{A_1}(r^2) = \big(C_p(r)\big)^2 = 1$. (Because ff_1 is not divisible by A_1, but A is divisible by A_1, $r^2 = ff_1 + A$ is not divisible by A_1, so $C_p(r) = \pm 1$.) Thus, $C_{A_1}(f)$ and $C_{A_1}(f_1)$ are either both 1 or both -1, so the modules $[f_i, g_i + \sqrt{A}]$ in the sequence of successors of $[f, g + \sqrt{A}]$ all satisfy $C_{A_1}(f) = C_{A_1}(f_i)$ as long as $C_{A_1}(f_i) \neq 0$.

But if $C_{A_1}(f_i) = 0$ for some i, then $C_{A_1}(f_{i-1}) = C_{A_1}(f_{i+1}) \neq 0$, as can be seen in the following way. When $f_{i-1}f_i = r_i^2 - A$ and $f_i f_{i+1} = r_{i+1}^2 - A$ are multiplied, one finds $f_{i-1}f_i^2 f_{i+1} = (r_i r_{i+1} + A)^2 - A(r_i + r_{i+1})^2$ by Brahmagupta's formula. Both r_i and r_{i+1} are divisible by A_1 and, as was seen above, f_i is divisible just once by A_1. Therefore, $f_{i-1} \cdot (\frac{f_i}{A_1})^2 \cdot f_{i+1} = (\frac{r_i r_{i+1} + A}{A_1})^2 - A \cdot (\frac{r_i + r_{i+1}}{A_1})^2$ is a square mod A_1 (because the right side is) that is not zero mod A_1 (because the left side is not). Since the same is true of $(\frac{f_i}{A_1})^2$, the desired conclusion $C_{A_1}(f_{i-1}f_{i+1}) = 1$ follows.

Thus, if $f \not\equiv 0 \bmod A_1$, then $C_{A_1}(f_i) = C_{A_1}(f)$ for all successors $[f_i, g_i + \sqrt{A}]$ of $[f, g + \sqrt{A}]$ in which $C_{A_1}(f_i) \neq 0$. If $[f, g + \sqrt{A}]$ is equivalent to $[f', g' + \sqrt{A}]$ and neither f nor f' is 0 mod A, then $C_{A_1}(f) = C_{A_1}(f')$ because, by the Theorem of Chapter 23, both are $C_{A_1}(f'')$ where $[f'', g'' + \sqrt{A}]$ is a stable successor of $[f, g + \sqrt{A}]$ in which $f'' \not\equiv 0 \bmod A_1$. □

For example, when $A = 105$, there are 8 equivalence classes of modules. The modules $[1, \sqrt{105}]$, $[105, \sqrt{105}]$, $[3, \sqrt{105}]$, $[15, \sqrt{105}]$, $[2, 1 + \sqrt{105}]$, $[4, 1 + \sqrt{105}]$, $[6, 3 + \sqrt{105}]$, and $[10, 5 + \sqrt{105}]$ represent the eight classes. For any $[f, g + \sqrt{A}]$ in the first equivalence class, $C_3(f) = C_5(f) = C_7(f) = 1$ for the three prime factors 3, 5, and 7 of 105 by Proposition 1, provided f is relatively prime to 105. This statement will be abbreviated by saying that *the signature of this first equivalence class is* $+ + +$. The second equivalence class contains the successor $[104, 1 + \sqrt{105}]$ of $[105, \sqrt{105}]$ so the values of $C_3(f)$, $C_5(f)$, and $C_7(f)$ for any $[f, g + \sqrt{A}]$ in this class for which f is relatively prime to 105 can be determined by determining $C_3(104) = C_3(2) = -1$, $C_5(104) = C_5(4) = 1$, and $C_7(104) = C_7(6) = -1$. Thus, the signature of this class is $-+-$. In similar ways, the signatures of the remaining six classes can be found to be $+--$, $--+$, $--+$, $+++$, $-+-$, and $+--$, respectively.

These observations go a long way toward establishing the truth of Euler's law in the case $A = 105$ because they show that if $C_{105}(p) = 1$ for some prime p, then, because there is a module $[p, k+\sqrt{A}]$ for some k, the quadratic characters of p with respect to the prime factors 3, 5, and 7 of 105 must be those of one of the 4 signatures found above. In particular, $C_{105}(11) = 1$ is impossible because if it were possible, there would be a module $[11, g + \sqrt{105}]$ in canonical form and the signature of this module would be $-++$ (because $C_3(11) = -1$, $C_5(11) = 1$, and $C_7(11) = 1$) which is not among the signatures that were found. For the same reason, $C_{105}(p) = 1$ is impossible whenever p is prime and congruent to 11 mod 105. In the same way, $C_{105}(p) = 1$ is impossible for exactly half of the possible congruence classes for p mod 105, namely, the classes mod 105 that are relatively prime to 105 that contain primes p for which the signs of $C_3(p)$, $C_5(p)$, and $C_7(p)$ follow one of the patterns $---$, $+-+$, $-++$, or $++-$.

In many cases, there is an *additional* condition that p must satisfy in order for $C_A(p) = 1$ to be possible. For example, when $A \equiv 3 \bmod 4$, there is a restriction on the *odd* values of f that occur in any given cycle:

Proposition 2. *Given a squarefree number A that is $3 \bmod 4$ and given a module $[f, g+\sqrt{A}]$, there is a module $[f', g'+\sqrt{A}]$ in canonical form equivalent to $[f, g + \sqrt{A}]$ in which f' is odd. The value of f' is the same mod 4 in all such modules $[f', g' + \sqrt{A}]$.*

Proof. The formula $ff_1 = r^2 - A$ shows that if f is even, then r is odd and $ff_1 \equiv 1 - A \equiv 2 \bmod 4$, so both $\frac{f}{2}$ and f_1 are odd. Thus, the successors of any module include modules in which f is odd. If both f and f_1 are odd, then r must be even and $ff_1 \equiv -A \equiv 1 \bmod 4$, from which $f \equiv f_1 \bmod 4$ follows. If f_i is even in some successor, then f_{i-1} and f_{i+1} are both odd, as is $\frac{f_i}{2}$, and the formula

$f_{i-1} \cdot (\frac{f_i}{2})^2 \cdot f_{i+1} = (\frac{r_i r_{i+1}+A}{2})^2 - A \cdot (\frac{r_i + r_{i+1}}{2})^2$ shows, because the number on the left is odd, that just one of the squares on the right is odd; since the square of an odd number is 1 mod 4 while the square of an even number is 0 mod 4, it follows that $f_{i-1}f_{i+1} \equiv 1 - A \cdot 0$ or $f_{i-1}f_{i+1} \equiv 0 - A \cdot 1 \bmod 4$, so $f_{i-1}f_{i+1} \equiv 1$ and $f_{i-1} \equiv f_{i+1} \bmod 4$ in either case. Thus, the value of f' mod 4 is the same in all successors $[f', g' + \sqrt{A}]$ of $[f, g + \sqrt{A}]$ in which f' is odd and the remaining statements of the proposition follow as before. □

For example, when $A = 195 = 3 \cdot 5 \cdot 13$, there are 8 cycles, of which the 8 modules $[1, \sqrt{195}]$, $[195, \sqrt{195}]$, $[3, \sqrt{195}]$, $[5, \sqrt{195}]$, $[13, \sqrt{195}]$, $[15, \sqrt{195}]$, $[39, \sqrt{195}]$, and $[65, \sqrt{195}]$ are representatives. Their signatures, as defined above are $+++$, $-++$, $+-+$, $-+-$, $+--$, $---$, $++-$, and $--+$, respectively, as is easily found. (For example $[3, \sqrt{195}] \sim [10, 5 + \sqrt{195}]$ and $C_3(10) = 1$ shows that the first sign of the third signature is $+$, while the remaining two signs come from $C_5(3) = -1$ and $C_{13}(3) = 1$.) Proposition 2 implies that a fourth sign can be annexed to the signature of each cycle. As is easily checked, in all eight cases the additional sign is determined by the condition that the total number of minus signs is always even. Thus, the eight signatures are $++++$, $-++-$, $+-+-$, $-+-+$, $+--+$, $----$, $++--$, and $--++$, respectively.

Since A is assumed to be squarefree, the even values of A to be considered are those that are $\equiv 2$ or $6 \bmod 8$. In each of these cases, a sign that can be annexed to the signature is determined by:

Proposition 3. *Given a squarefree number A that is $2 \bmod 4$ and given a module $[f, g+\sqrt{A}]$, there is a module $[f', g'+\sqrt{A}]$ in canonical form equivalent to $[f, g+\sqrt{A}]$ in which f' is odd. When $A \equiv 2 \bmod 8$, two such values of f', call them f' and f'', must satisfy either $f' \equiv f'' \bmod 8$ or $f' \equiv -f'' \bmod 8$. When $A \equiv 6 \bmod 8$, they must satisfy either $f' \equiv f'' \bmod 8$ or $f' \equiv 3f'' \bmod 8$.*

Proof. As before, the key formula is $ff_1 = r^2 - A$. If f is even, then r must be even, which implies that $ff_1 \equiv -A \equiv 2 \bmod 4$, so $\frac{f}{2}$ and f_1 must be odd, which proves the first statement of the theorem. If f and f_1 are both odd, then $r^2 - A$ and r are odd, so $ff_1 \equiv 1 -$

$A \bmod 8$; thus, $f_1 \equiv -f \bmod 8$ when $A \equiv 2 \bmod 8$ and $f_1 \equiv 3f \bmod 8$ when $A \equiv 6 \bmod 8$. Finally, if f_i is even, then $f_{i-1} \cdot (\frac{f_i}{2})^2 \cdot f_{i+1} = (\frac{r_i r_{i+1}+A}{2})^2 - A \cdot (\frac{r_i + r_{i+1}}{2})^2$ is odd, which implies that $\frac{r_i r_{i+1}+A}{2}$ is odd, from which $f_{i-1}f_{i+1} \equiv 1$ or $1 - A \bmod 8$, depending on whether $\frac{r_i+r_{i+1}}{2}$ is even or odd, and the proposition follows as before. □

For example, when $A = 30$, the signature of an equivalence class of modules can be taken to contain three signs, the first two signs, of $C_3(f)$ and $C_5(f)$, determined as above and a third sign which is + when $f \equiv 1$ or $3 \bmod 8$ and minus when $f \equiv 5$ or $7 \bmod 8$ for an odd f. There are four equivalence classes of modules when $A = 30$ represented by, for example, $[1, \sqrt{30}]$, $[30, \sqrt{30}]$, $[2, \sqrt{30}]$, and $[10, \sqrt{30}]$. Their signatures are easily found to be $+++$, $-+-$, $--+$, and $+--$, respectively.

Definition. *For a given squarefree number A, the* **signature** *of an equivalence class of modules for that A is a sequence of signs, + or −, determined in the following way. Let $A_1, A_2, \ldots, A_m$ be the odd prime factors of A ordered by $A_1 < A_2 < \cdots < A_m$. (If $A = 2$, then $m = 0$; otherwise $m > 0$.) The ith sign of the signature for $i \leq m$ is found by finding a module $[f, g + \sqrt{A}]$ in the equivalence class for which $f \not\equiv 0 \bmod A_i$ and taking the ith sign to be the sign of $C_{A_i}(f)$ for this f. If $A \equiv 1 \bmod 4$, the m signs determined in this way are the complete signature. Otherwise, there is one more sign, an $(m+1)$st sign, determined by finding a module $[f, g + \sqrt{A}]$ in the equivalence class in which f is odd and determining that last sign according to the rules:*

if $A \equiv 3 \bmod 4$, the sign is + if $f \equiv 1 \bmod 4$ and − otherwise,

if $A \equiv 2 \bmod 8$, the sign is + if $f \equiv \pm 1 \bmod 4$ and − otherwise,

if $A \equiv 6 \bmod 8$, the sign is + if $f \equiv 1$ or $3 \bmod 8$ and − otherwise.

Notation: Let λ_1, λ_2, and λ_3 denote the functions that assign ± 1 to odd numbers that give the last sign of the signatures for A in the cases $A \equiv 3 \bmod 4$, $A \equiv 2 \bmod 8$, and $A \equiv 6 \bmod 8$, respectively. That is, for an odd number f, $\lambda_1(f)$ is 1 if $f \equiv 1 \bmod 4$, -1 if $f \equiv 3 \bmod 4$, while $\lambda_2(f)$ is 1 if $f \equiv \pm 1 \bmod 8$, -1 if $f \equiv \pm 3 \bmod 8$, and $\lambda_3(f)$ is 1 if $f \equiv 1$ or $3 \bmod 8$, -1 if $f \equiv 5$ or $7 \bmod 8$. (Note that $\lambda_3(f) = \lambda_1(f)\lambda_2(f)$.)

Exercises for Chapter 24

1. Verify the characters given for several of the cycles for square-free A given in the table of stable modules in the appendix.

2. Verify several cases of the fact that *the signature of a product is the product of the signatures* of two primitive cycles (where the product of the signatures is defined in an obvious way).

Chapter 25

The Main Theorem

Let A be a given squarefree number. The signature of a given equivalence class of modules for A is defined in the last chapter as a sequence of signs $+$ or $-$; when m is the number of odd prime factors of A, there are m signs when $A \equiv 1 \bmod 4$ and $m+1$ signs otherwise. To find the signature of an equivalence class, it suffices[1] to find one module $[f, g + \sqrt{A}]$ in the class in which f is odd and relatively prime to A. Then the first m signs of the signature are $C_{A_i}(f)$, where A_i is the ith odd prime factor of A (ordered by $A_i < A_{i+1}$), and the last sign, in case $A \not\equiv 1 \bmod 4$, is $\lambda_1(f)$, $\lambda_2(f)$, or $\lambda_3(f)$, when $A \equiv 3 \bmod 4$, $\equiv 2 \bmod 8$, or $\equiv 6 \bmod 8$, respectively. (See the definitions of the $\lambda_i(f)$ at the end of Chapter 24.)

It is natural to define the **signature relative to** A of an odd number f that is relatively prime to A to be this same sequence of signs—m signs $C_{A_i}(f)$ when $A \equiv 1 \bmod 4$, and these m signs followed by $\lambda_1(f)$, $\lambda_2(f)$, or $\lambda_3(f)$ when $A \equiv 3 \bmod 4$, $\equiv 2 \bmod 8$, or $\equiv 6 \bmod 8$, respectively.

Main Theorem. *Let A be squarefree and let p be an odd prime that does not divide A. Then $C_p(A)$ is the product of the signs of the*

[1]The definition in Chapter 24 allows for the use of different f's for the determination of different signs in the signature. As will be proved in Chapter 27, a single f can be used as above to determine all the signs provided the class is *primitive*.

signature of p for this A. In other words, A is a square mod p if and only if the signature of p relative to A contains an even number of minus signs.

Corollary (Euler's law). *The value of $C_p(A)$ depends only on the value of p mod $4A$.*

Deduction. As was explained at the beginning of Chapter 24, Euler's law for *squarefree* A implies Euler's law for all A. Assume, therefore, that A is squarefree. By the theorem, $C_p(A)$ depends only on the signs in the signature of p relative to A. The ith sign $C_{A_i}(p)$ for $i = 1, 2, \dots, m$ depends only on the value of p mod A_i and therefore depends only on the value of p mod $4A$. The $(m+1)$st sign, if there is one, depends only on the value of p mod 4 if $A \equiv 3 \bmod 4$ and only on the value of p mod 8 if $A \equiv 2$ or $6 \bmod 8$. Since 4 divides $4A$ in the first case and 8 divides $4A$ in the last two cases, the corollary follows. □

The Main Theorem will be proved in Chapter 29.

Exercise for Chapter 25

Check that for squarefree values of A and for primitive modules, the signatures shown in the appendix are precisely those in which there are an even number of minuses. (Because $C_p(A) = 1$ if and only if $A \not\equiv 0 \bmod p$ and $[p, g + \sqrt{A}]$ is in canonical form for some g, this observation is equivalent to the Main Theorem for the squarefree values of A included in the appendix.)

Chapter 26

Modules That Become Principal When Squared

This chapter finds all solutions $[f, g+\sqrt{A}]$ of the problem $[f, g+\sqrt{A}]^2 \sim [1]$ in all cases in which A is prime or A is a product of two primes, each congruent to $3 \bmod 4$. The result will be used in the following three chapters to prove the law of quadratic reciprocity and the Main Theorem.

Of course, $[1]^2 = [1]$ is a solution of this problem. Another simple solution is $[A, \sqrt{A}]^2 = [A^2, A\sqrt{A}, A] = [A][A, \sqrt{A}, 1] = [A] \sim [1]$. Any module that is equivalent to one of these solutions $[1]$ and $[A, \sqrt{A}]$ of the problem is also a solution, because the equivalence class of a product depends only on the equivalence classes of the factors.

As was seen in Chapter 23, it is natural to identify equivalence classes of modules with cycles of stable modules. The square of an equivalence class then becomes the square of the corresponding cycle of stable modules, which is found by choosing a module in the cycle, squaring it, and finding the cycle of stable modules equivalent to the result. The question then becomes: *For which cycles of stable modules is the "square" found in this way the principal cycle?*

Proposition. *If A is a prime congruent to* $1 \bmod 4$, *then the principal cycle is the only cycle whose square is the principal cycle.* (In this case, therefore, $[A, \sqrt{A}]$ must be in the principal cycle.) *If A is a prime congruent to* $3 \bmod 4$, *or if $A = pq$ where p and q are primes that are congruent to* $3 \bmod 4$, *then* $[1]$ *and* $[A, \sqrt{A}]$ *are in different cycles, and these two cycles are the only ones whose squares are the principal cycle.*

Lemma. *If $[f, g + \sqrt{A}]$ is a stable module and $[f_1, g_1 + \sqrt{A}]$ is its successor, the conjugate*[1] *of $[f_1, g_1 + \sqrt{A}]$ is stable, and the conjugate of $[f, g + \sqrt{A}]$ is its successor.*

Proof. Suppose that $[f, g + \sqrt{A}]$ and $[f_1, g_1 + \sqrt{A}]$ are both in canonical form, so that their conjugates are $[f, g' + \sqrt{A}]$ and $[f_1, g_1' + \sqrt{A}]$, respectively, where g' is the least solution of $g' + g \equiv 0 \bmod f$ and g_1' is the least solution of $g_1' + g_1 \equiv 0 \bmod f_1$. By definition, $g_1 \equiv r \bmod f_1$ where r is the number used by the comparison algorithm to determine the successor of $[f, g + \sqrt{A}]$. Therefore, $r + g_1' \equiv 0 \bmod f_1$, so r is the number used by the comparison algorithm to determine the successor of $[f_1, g_1' + \sqrt{A}]$ if and only if no number r_1 satisfies $r_1 < r$, $r_1 \equiv r \bmod f_1$, and $r_1^2 > A$. That there is no such r_1 follows from the assumption that $[f, g + \sqrt{A}]$ is stable, because this assumption means $|r - f|^2 < A$, and therefore implies, in succession, $r^2 + f^2 < A + 2rf$, $ff_1 + f^2 < 2rf$, $f_1 + f < 2r$, $f_1^2 + ff_1 < 2rf_1$, $f_1^2 + r^2 < 2rf_1 + A$, and $|f_1 - r|^2 < A$; this last inequality shows there is no r_1 as above, because $r_1 < r$ and $r_1 \equiv r \bmod f_1$ imply that $r_1 + f_1 \le r$ and therefore imply $r_1^2 \le (r - f_1)^2 < A$.

The inequality $|f_1 - r|^2 < A$ then implies that $[f_1, g_1' + \sqrt{A}]$ is stable.

Finally, because r is the number used by the comparison algorithm to determine the successor, call it $[F, G + \sqrt{A}]$, of $[f_1, g_1' + \sqrt{A}]$, $Ff_1 = r^2 - A$ holds. Since this number is also ff_1, F must be f. Finally, $G \equiv r \bmod F$ then means $G \equiv r \bmod f$, so $G + g \equiv r + g \equiv 0 \bmod f$ and $G \equiv g' \bmod f$, as was to be shown. □

[1]See Chapter 20 for the definition of the conjugate of a module.

Simply put, the lemma states that the conjugate of a cycle is a cycle, but conjugation reverses the order in which cycles are traversed.

Proof of the proposition. If $[f, g + \sqrt{A}]^2 \sim [1]$, then $[f, g + \sqrt{A}]$ is primitive, as follows directly from the proposition of Chapter 23. (The equivalence class of $[f, g+\sqrt{A}]$ times *itself* is the principal class.) Therefore, it suffices to find the cycles of *primitive* modules whose squares are principal.

If $[f, g+\sqrt{A}]$ is stable and primitive and if $[f, g+\sqrt{A}]^2 \sim [1]$, then $[f, g' + \sqrt{A}] \sim [f, g' + \sqrt{A}][f, g + \sqrt{A}]^2 \sim [1][f, g + \sqrt{A}] \sim [f, g + \sqrt{A}]$ when $[f, g' + \sqrt{A}]$ is the conjugate of $[f, g + \sqrt{A}]$. Therefore, by the lemma and by the theorem of Chapter 23, $[f, g + \sqrt{A}]$ and its conjugate are in the same cycle, say $[f, g' + \sqrt{A}] = [f_i, g_i + \sqrt{A}]$ where $[f_i, g_i + \sqrt{A}]$ is the ith successor of $[f, g + \sqrt{A}]$. The lemma then implies, when $i \geq 1$, that $[f_{i-1}, g_{i-1} + \sqrt{A}]$ is the conjugate of $[f_1, g_1 + \sqrt{A}]$ because both are the stable module whose successor is the conjugate of $[f, g+\sqrt{A}]$. In the same way, $[f_{i-2}, g_{i-2}+\sqrt{A}]$ is the conjugate of $[f_2, g_2+\sqrt{A}]$ when $i \geq 2$, $[f_{i-3}, g_{i-3}+\sqrt{A}]$ the conjugate of $[f_3, g_3+\sqrt{A}]$ when $i \geq 3$, and so forth. If i is even, say $i = 2j$, then $[f_j, g_j + \sqrt{A}]$ is its own conjugate. If i is odd, say $i = 2j + 1$, then $[f_j, g_j+\sqrt{A}]$ is the conjugate of $[f_{j+1}, g_{j+1}+\sqrt{A}]$. Let a stable module $[f, g+\sqrt{A}]$ be called **pivotal of type 1** if it is its own conjugate and **pivotal of type 2** if it is the conjugate of its successor. Thus, a cycle whose square is equivalent to $[1]$ must contain a pivotal module. Since the converse is clear, the proposition amounts to a description of the primitive cycles that contain pivotal modules, and it can be proved by finding all pivotal modules.

When A is an odd prime, there are just four pivotal modules. They can be determined in the following way.

If $[f, g+\sqrt{A}]$ is a pivotal module of type 1—that is, $[f, g+\sqrt{A}] = [f, g' + \sqrt{A}]$—then $g + g \equiv 0 \bmod f$ when $[f, g + \sqrt{A}]$ is in canonical form. Because $A \equiv g^2 \bmod f$, the condition $2g \equiv 0 \bmod f$ implies $4A \equiv (2g)^2 \equiv 0 \bmod f$, so f must divide $4A$. Since A is an odd prime, f must therefore have one of the values 1, 2, 4, A, $2A$, or $4A$. Since $[f, g + \sqrt{A}]$ is stable, $f \leq A$, so f can only be 1, 2, 4, or A. The value $f = 4$ is impossible because $2g \equiv 0 \bmod 4$ would imply g

was even, which is contrary to the assumption that A is odd because the other assumptions imply $A \equiv g^2 \bmod 4$. Each of the remaining values 1, 2, and A indeed gives rise to a pivotal module of type 1, namely, $[1, \sqrt{A}]$, $[2, 1+\sqrt{A}]$, and $[A, \sqrt{A}]$, respectively.

The pivotal modules of type 2 when A is an odd prime are even easier to find. If $[f, g+\sqrt{A}]$ is in canonical form and pivotal of type 2, then $f = f_1$, which implies that $f^2 = r^2 - A$, where r is the number used by the comparison algorithm to find the successor $[f_1, g_1 + \sqrt{A}]$ of $[f, g_1+\sqrt{A}]$. Then $A = r^2 - f^2 = (r+f)(r-f)$, so the existence of a pivotal module of type 2 implies a factorization of A into two factors. Since A is prime, it follows that $r - f = 1$ and $r + f = A$, which is to say that $r = \frac{A+1}{2}$ and $f = f_1 = \frac{A-1}{2}$. Because $g_1 \equiv r \bmod f_1$, one then finds $g_1 = 1$ and $g = f - g_1 = \frac{A-3}{2}$. In fact, $[\frac{A-1}{2}, \frac{A-3}{2} + \sqrt{A}]$ is easily seen to be pivotal of type 2. (The value of r used to go from it to its successor is $\frac{A+1}{2}$ because $\frac{A+1}{2} + \frac{A-3}{2} \equiv 0 \bmod \frac{A-1}{2}$ and $(\frac{A+1}{2})^2 > A$ but the square of $\frac{A+1}{2} - \frac{A-1}{2} = 1$ is not greater than A.) Thus, $[\frac{A-1}{2}, \frac{A-3}{2} + \sqrt{A}]$ is the unique pivotal module of type 2 in this case.

A cycle which contains a pivotal module must in fact contain *two* of them, unless it consists of only one module, as can be seen in the following way. Let the modules in a cycle be numbered starting with a pivotal module $[f_0, g_0 + \sqrt{A}]$. If $[f_0, g_0 + \sqrt{A}]$ is of type 1, then $[f_i, g_i + \sqrt{A}]$ is the conjugate of $[f_{l-i}, g_{l-i} + \sqrt{A}]$ for each i, where l is the length of the cycle. If l is even, say $l = 2j$, then $[f_j, g_j + \sqrt{A}]$ is pivotal of type 1; if l is odd, say $l = 2j+1$, then $[f_j, g_j + \sqrt{A}]$ is pivotal of type 2. Thus, the cycle contains another pivotal module *unless* $l = 1$, in which case $[f_0, g_0 + \sqrt{A}]$ is pivotal of type 2 as well as type 1. If $[f_0, g_0 + \sqrt{A}]$ is pivotal of type 2, then $[f_i, g_i + \sqrt{A}]$ is the conjugate of $[f_{l-i+1}, g_{l-i+1} + \sqrt{A}]$ for each i, from which it follows in a similar way that when $l = 2j+1$ is odd, $[f_j, g_j + \sqrt{A}]$ is pivotal of type 1 and when $l = 2j$ is even, $[f_j, g_j + \sqrt{A}]$ is pivotal of type 2. Thus, there is a second pivotal module in the cycle except when $l = 1$, in which case the sole module in the cycle is pivotal of both types.

The only cases in which the module $[\frac{A-1}{2}, \frac{A-3}{2} + \sqrt{A}]$ pivotal of type 2 is also pivotal of type 1 are those in which $\frac{A-1}{2} = 1$ or 2, which is to say the cases $A = 3$ or 5.

In summary, when A is an odd prime, only two cycles of stable modules contain pivotal modules, namely, the cycles of the modules $[1, \sqrt{A}]$, $[2, 1 + \sqrt{A}]$, $[A, \sqrt{A}]$, and $[\frac{A-1}{2}, 1 + \sqrt{A}]$. These modules lie in exactly two cycles; except when $A = 3$ or 5, there are four pivotal modules, two in each cycle, but in the exceptional cases, there are only three pivotal modules and one of them is a cycle unto itself.

When A is prime and $A \equiv 1 \bmod 4$, the module $[2, 1 + \sqrt{A}]$ is not primitive, so the square of its cycle cannot be the principal cycle. Therefore, for prime values of A that are 1 mod 4 the principal cycle is the only one whose square is the principal cycle, as was to be shown.

When A is prime and $A \equiv 3 \bmod 4$, all four of the pivotal modules are primitive, as is easily checked. Therefore, two cycles solve the problem. Moreover, the signature of $[A, \sqrt{A}]$ is $--$, as is easily seen. (The second sign is $-$ directly from its definition; the first sign is $-$ because the successor of $[A, \sqrt{A}]$ is $[A-1, 1+\sqrt{A}]$ and $(A-1)^{(A-1)/2} \equiv (-1)^{(A-1)/2} \equiv -1 \bmod A$ when $A \equiv 3 \bmod 4$.) Therefore, the cycle of $[A, \sqrt{A}]$ is not the principal cycle and the proposition follows in this case.

Consider finally the case in which $A = pq$, for distinct primes p and q that are both 3 mod 4. Again, the pivotal modules of type 1 must have the form $[f, g + \sqrt{A}]$ where f divides $4A$. Again, f is at most A and cannot be divisible by 4. Thus, f has one of the values 1, 2, p, $2p$, q, $2q$, pq. It is easily checked that the modules $[1, \sqrt{A}]$, $[2, 1 + \sqrt{A}]$, $[p, \sqrt{A}]$, $[2p, p + \sqrt{A}]$, $[q, \sqrt{A}]$, $[2q, q + \sqrt{A}]$, and $[pq, \sqrt{A}]$ are all pivotal of type 1 *except* that $[2p, p + \sqrt{A}]$ is not stable when p is the larger of the two factors of A, because the r that determines the successor of $[2p, p+\sqrt{A}]$ is p, and $|p - 2p|^2 = p^2 > pq = A$. Thus, there are exactly 6 pivotal modules of type 1 in this case.

As for the pivotal modules of type 2, again $[\frac{A-1}{2}, \frac{A-3}{2} + \sqrt{A}]$ is such a module. A second one corresponds to the factorization $r - f = q$, $r + f = p$ of $A = r^2 - f^2$, where p is the larger of the two primes. Then $f = \frac{p-q}{2}$ and $r = \frac{p+q}{2}$, which gives $[f, r + \sqrt{A}] = [\frac{p-q}{2}, q + \sqrt{A}]$

as the successor of a pivotal module of type 2 and therefore gives $[\frac{p-q}{2}, \frac{p-q}{2} - q + \sqrt{A}] = [\frac{p-q}{2}, \frac{p-3q}{2} + \sqrt{A}]$ as the sole remaining pivotal module of type 2 in this case.

As before, the eight pivotal modules lie in four cycles, each of which contains two pivotal modules, except when two of them coincide (which happens just when $p = q+4$ so that the module $[\frac{p-q}{2}, q+\sqrt{A}]$ is $[2, 1+\sqrt{A}]$) in which case there are still 4 cycles but one of them contains only one module.

In this case (when $A = pq$, where $p > q$ are both prime and congruent to 3 mod 4), A is 1 mod 4, so $[2, 1+\sqrt{A}]$ and $[\frac{A-1}{2}, 1+\sqrt{A}]$ are not primitive, and neither are $[2q, q+\sqrt{A}]$ or $[\frac{p-q}{2}, q+\sqrt{A}]$. The remaining four, $[1, \sqrt{A}]$, $[p, \sqrt{A}]$, $[q, \sqrt{A}]$, and $[A, \sqrt{A}]$, are all primitive and they determine two cycles whose squares are the principal cycle. One is the principal cycle and the other is the cycle of $[A, \sqrt{A}]$, because the signature of $[A, \sqrt{A}]$ is the signature of its successor $[A-1, 1+\sqrt{A}]$, which is $--$ because $A-1$ is $-1 \bmod p$ and $-1 \bmod q$, neither of which is a square,[2] which completes the proof of the proposition. □

Exercises for Chapter 26

1. Most modules in the table in the appendix are equivalent to pivotal modules. Find at least one that is not.

2. Find, if possible, modules that have the same signature but are not equivalent.

[2]The index of -1 for a primitive root $g \bmod p$ is $(p-1)/2$, which is odd, so -1 is not a square mod p and, for the same reason, -1 is not a square mod q.

Chapter 27

The Possible Signatures for Certain Values of A

The proposition of the last chapter implies the following theorem, which in turn implies the law of quadratic reciprocity as it is stated and proved in Chapter 28. As before, all hypernumbers and modules relate to a fixed A, not a square.

Theorem. *If A is a prime congruent to* $1 \bmod 4$, *then all primitive cycles of stable modules have signature* $+$. *If A is a prime congruent to* $3 \bmod 4$, *then half of the primitive cycles of stable modules have signature* $++$ *and the other half have signature* $--$. *When A is a product $A = pq$ of two primes p and q, both congruent to* $3 \bmod 4$, *half of the primitive cycles of stable modules have signature* $++$ *and the other half have signature* $--$*; moreover,* $[p, \sqrt{pq}]$ *and* $[q, \sqrt{pq}]$ *are in opposite halves.*

Proof. Let A be a prime congruent to $1 \bmod 4$ and consider the *squaring* function which assigns to each primitive cycle of stable modules its square, as in the proposition of the last chapter. If the cycles of $[f, g + \sqrt{A}]^2$ and $[F, G + \sqrt{A}]^2$ are the same, then the cycle of $[f, g' + \sqrt{A}][F, G + \sqrt{A}]$, where $[f, g' + \sqrt{A}]$ is the conjugate of $[f, g + \sqrt{A}]$, has the cycle of $[1]$ as its square (because the square of a product is the product of the squares and the product of $[f, g' + \sqrt{A}]^2$ and $[F, G + \sqrt{A}]^2$ is equivalent to the product of $[f, g' + \sqrt{A}]^2$ and

$[f, g+\sqrt{A}]^2$, which is equivalent to $[1]^2 = [1]$ by virtue of the assumption that $[f, g+\sqrt{A}]$ is primitive). Therefore, by the proposition of the last chapter, $[f, g'+\sqrt{A}][F, G+\sqrt{A}]$, having its square equivalent to [1], must itself be equivalent to [1], which is to say that $[f, g+\sqrt{A}]$ and $[F, G+\sqrt{A}]$ are in the same cycle (multiply both sides of $[f, g'+\sqrt{A}][F, G+\sqrt{A}] \sim [1]$ by $[f, g+\sqrt{A}]$). In short, the squaring function on primitive cycles is one-to-one. Since a one-to-one function from a finite set to itself must also be onto, it follows that *every cycle of primitive stable modules when* $A \equiv 1 \bmod 4$ *is the square of some other cycle of primitive stable modules.*

The lemma below states that the signature of a product of two modules is the product of the signatures. Therefore, the signature of any primitive module is the square of the signature of some other primitive module, which means that its signature must be $+$.

Next let A be a prime congruent to 3 mod 4. As before, $[f, g+\sqrt{A}]^2 \sim [F, G+\sqrt{A}]^2$ implies that the square of $[f, g'+\sqrt{A}][F, G+\sqrt{A}]$ is equivalent to [1], which means in this case that $[f, g'+\sqrt{A}][F, G+\sqrt{A}]$ is equivalent either to [1] or to $[A, \sqrt{A}]$. Therefore, either $[F, G+\sqrt{A}] \sim [f, g+\sqrt{A}]$ or $[F, G+\sqrt{A}] \sim [f, g+\sqrt{A}][A, \sqrt{A}]$. Since, as was seen in Chapter 26, [1] and $[A, \sqrt{A}]$ are in different cycles, the squaring function in this case is therefore a *two-to-one* function from primitive cycles to themselves, and exactly half of the primitive cycles are squares of primitive cycles; thus, half of the primitive cycles have the signature $++$. Since the signature of $[A, \sqrt{A}]$ is $--$ (because $\lambda_1(A) = -1$ and $[A-1, 1+\sqrt{A}]$ is the successor of $[A, \sqrt{A}]$ and $C_A(A-1) = -1$), the cycle of $[A, \sqrt{A}]$ times any cycle that is a square has signature $--$, which accounts for the other half of the cycles and shows that they all have signature $--$.

Finally, when $A = pq$ where p and q are primes for which $p \equiv q \equiv 3 \bmod 4$, the fact that just two cycles have square the principal cycle implies in the same way that the squaring function from primitive cycles to primitive cycles is two-to-one, so half of the primitive cycles are squares and therefore have signature $++$. The cycle of $[A, \sqrt{A}]$ again has signature $--$ (both $C_p(A-1)$ and $C_q(A-1)$ are -1), so an equal number of cycles have the signature $--$, which accounts for all of the primitive cycles. Finally, the four primitive pivotal modules

$[1]$, $[p, \sqrt{pq}]$, $[q, \sqrt{pq}]$, and $[A, \sqrt{A}]$ are in two different cycles, with two of them in each of the two; since $[1]$ and $[A, \sqrt{A}]$ lie in different cycles, so must $[p, \sqrt{pq}]$ and $[q, \sqrt{pq}]$. □

Lemma. *If $[f, g + \sqrt{A}]$ and $[F, G + \sqrt{A}]$ are primitive modules in canonical form, then the signature of their product is the product of their signatures.*

(Signatures are of course multiplied by multiplying corresponding signs using the rule that the product of like signs is $+$ and the product of differing signs is $-$.)

Proof. If f and F are relatively prime to $4A$, then the signatures of $[f, g + \sqrt{A}]$ and $[F, G + \sqrt{A}]$ are simply the signatures of f and F, respectively, in the obvious sense. If f and F are also *relatively prime to each other*, the signature of the product $[f, g + \sqrt{A}][F, G + \sqrt{A}]$ is the signature of fF, because when the Chinese remainder theorem is used to find a solution $\mathcal{G}$ of $\mathcal{G} \equiv g \bmod f$ and $\mathcal{G} \equiv G \bmod F$, one finds $[f, g+\sqrt{A}][F, G+\sqrt{A}] = [f, \mathcal{G}+\sqrt{A}][F, \mathcal{G}+\sqrt{A}] = [fF, f\cdot(\mathcal{G}+\sqrt{A}), F\cdot(\mathcal{G}+\sqrt{A}), (\mathcal{G}+\sqrt{A})^2] = [fF, \mathcal{G}+\sqrt{A}]$, because the existence of a solution (a, b) of $af = bF + 1$ makes it possible to annex $\mathcal{G}+\sqrt{A}$ to the list and then to drop $f\cdot(\mathcal{G}+\sqrt{A})$, $F\cdot(\mathcal{G}+\sqrt{A})$, and $(\mathcal{G}+\sqrt{A})^2$.

Because each individual sign of the signature is multiplicative—for each prime p that does not divide fF, $C_p(fF) = C_p(f)C_p(F)$ and similarly for λ_1, λ_2, and λ_3 when fF is odd—these simple observations suffice to prove the lemma once it is shown that the following construction—which makes it possible to replace $[f, g+\sqrt{A}]$ with an equivalent module in which f is relatively prime to $4A$ and replace $[F, G+\sqrt{A}]$ with an equivalent module in which F is relatively prime to both $4A$ and f—is possible:

Construction. *Given a primitive module and a number N, construct an equivalent module in whose canonical form $[\mathcal{F}, \mathcal{G}+\sqrt{A}]$ the number $\mathcal{F}$ is relatively prime to N.*

Let $[f, g+\sqrt{A}]$ be the canonical form of the given primitive module, which can be assumed without loss of generality to have $e = 1$ because $[e][f, g + \sqrt{A}]$ is equivalent to $[f, g + \sqrt{A}]$. Let r be the number used by the comparison algorithm to find the successor of $[f, g + \sqrt{A}]$. The construction will be done in two steps; first it will be shown that there is a number T for which $\frac{(r+Tf)^2-A}{f}$, call it $Q(T)$, is relatively prime to fN, and then it will be shown that the module $[Q(T), r + Tf + \sqrt{A}]$, whose canonical form is $[Q(T), \mathcal{G} + \sqrt{A}]$ where $\mathcal{G}$ is the smallest solution of $\mathcal{G} \equiv r + Tf \bmod Q(T)$, is equivalent to $[f, g + \sqrt{A}]$.

For any given number t, let

$$Q(t) = \frac{(r + tf)^2 - A}{f} = t^2 f + 2rt + \frac{r^2 - A}{f}.$$

Thus, $Q(t)$ is a polynomial in t with number coefficients; it is to be shown that a number T can be found for which $Q(T)$ is relatively prime to fN.

For any given prime p, a number t_p can be found for which $Q(t_p) \not\equiv 0 \bmod p$. If the leading coefficient f of $Q(t)$ is not zero mod p, then $Q(t) \equiv 0 \bmod p$ has at most two roots; therefore, if $p > 2$, there is at least one t for which $Q(t) \not\equiv 0 \bmod p$. The case $p = 2$ will be treated last. If the leading coefficient is zero mod p but the second coefficient $2r$ is not zero mod p, then $Q(t) \equiv 0 \bmod p$ has just one root, so there are $p - 1$ solutions t of $Q(t) \not\equiv 0 \bmod p$. If both of these coefficients f and $2r$ are zero mod p, then the third coefficient $\frac{r^2-A}{f}$ is nonzero mod p by virtue of the assumption that $[f, g + \sqrt{A}]$ is primitive, in which case $t = 0$ has the required property. Finally, when $p = 2$, the primitivity of $[f, g + \sqrt{A}]$ implies that at least one of f and $\frac{r^2-A}{f}$ is odd; if $\frac{r^2-A}{f}$ is odd, then any even t has the required property and in the remaining case any odd t does.

Let a number t_p for which $Q(t_p) \not\equiv 0 \bmod p$ be chosen for each prime factor p of fN. By the Chinese remainder theorem, there is a T that satisfies all of the congruences $T \equiv t_p \bmod p$ simultaneously. Then $Q(T) = \frac{(r+Tf)^2-A}{f} \not\equiv 0 \bmod p$ for each prime factor p of fN. In short, $Q(T)$ is relatively prime to fN.

Finally, let $v = Tf + r$ for this T and consider the principal module $[v + \sqrt{A}] = [v^2 - A, v + \sqrt{A}] = [f \cdot Q(T), v + \sqrt{A}]$. Because f and $Q(T)$ are relatively prime, the proof given above shows that $[v + \sqrt{A}] = [f, v + \sqrt{A}][Q(T), v + \sqrt{A}]$. Since $[v + \sqrt{A}]$ is principal and the canonical forms of the two factors on the right are $[f, r + \sqrt{A}] = [f, g' + \sqrt{A}]$ and $[Q(T), \mathcal{G} + \sqrt{A}]$ where g' and $\mathcal{G}$ are the smallest solutions of $g' + g \equiv 0 \bmod f$ and $\mathcal{G} \equiv v \bmod Q(T)$, respectively, it follows that $[1] \sim [f, g' + \sqrt{A}][Q(T), \mathcal{G} + \sqrt{A}]$, and an equivalence $[f, g + \sqrt{A}] \sim [Q(T), \mathcal{G} + \sqrt{A}]$ of the required form follows when one multiplies by $[f, g + \sqrt{A}]$. □

Exercises for Chapter 27

1. If A is a prime that is 1 mod 4, then all primitive modules have signature $+$. Find such values of A for which there is more than one primitive cycle to find primitive modules that have the same signature but are not equivalent.

2. If A is a prime that is 3 mod 4, then all primitive modules have signature either $++$ or $--$. For such A find primitive modules that have the same signature but are not equivalent.

Chapter 28

The Law of Quadratic Reciprocity

The Law of Quadratic Reciprocity. *If p and q are distinct odd primes, and if either of them is* 1 mod 4, *then $C_p(q) = C_q(p)$, but if both are* 3 mod 4, *then $C_p(q) = -C_q(p)$.*

This law can be deduced from the theorem of Chapter 27 in the following way.

Proof. The case $p \equiv q \equiv 3 \bmod 4$ follows immediately from the theorem because $C_p(q)$ is the first sign of the signature of $[q, \sqrt{pq}]$ and $C_q(p)$ is the second sign of the signature of $[p, \sqrt{pq}]$ (when $p > q$), and the theorem states that one of these signatures is $++$ and the other is $--$.

If $p \equiv q \equiv 1 \bmod 4$ and if $C_p(q) = 1$, then there is a module whose canonical form is $[p, g + \sqrt{q}]$. The signature of this module, which is $C_q(p)$, is $+$ by the theorem. In short, $C_p(q) = 1$ implies $C_q(p) = 1$. By symmetry, $C_q(p) = 1$ implies $C_p(q) = 1$, or, what is the same, $C_p(q) = -1$ implies $C_q(p) = -1$. Therefore, $C_p(q) = C_q(p)$.

Finally, suppose $p \equiv 1 \bmod 4$ and $q \equiv 3 \bmod 4$. If $C_p(q) = 1$ there is a module whose canonical form is $[p, g + \sqrt{q}]$. The signature of this module is $++$ or $--$ by the theorem. Since $\lambda_1(p) = 1$, the second sign is $+$, so the first sign must also be $+$, which is to say

$C_q(p) = 1$. In short, $C_p(q) = 1$ implies $C_q(p) = 1$. If $C_q(p) = 1$, there is a module whose canonical form is $[q, g+\sqrt{p}]$ whose signature $C_p(q)$ is $+$. Thus $C_p(q) = -1$ implies $C_q(p) = -1$, and the proof is complete. □

The following proposition is often called a "Supplementary Law of Quadratic Reciprocity" because it gives the value of $C_p(2)$ and shows that this value depends only on the value of $p \bmod 8$. (The other "Supplementary Law of Quadratic Reciprocity" is the formula $C_p(-1) = \lambda_1(p)$ that follows from $C_p(-1) = (-1)^{(p-1)/2}$—Euler's criterion—for odd primes p. Note that neither statement expresses any kind of "reciprocity.")

Proposition. *For any odd prime p, $C_p(2) = \lambda_2(p)$.*

(By definition, $\lambda_2(n)$ is 1 for $n \equiv \pm 1 \bmod 8$ and -1 for $n \equiv \pm 3 \bmod 8$. See the end of Chapter 24.)

Proof. If $C_p(2) = 1$, there is a module whose canonical form is $[p, g+\sqrt{2}]$. When $A = 2$, the principal cycle is the only cycle of stable modules and its signature is $+$. Since the signature of $[p, g+\sqrt{2}]$ is $\lambda_2(p)$, it follows that $C_p(2) = 1$ implies $\lambda_2(p) = 1$.

It remains to show that $C_p(2) = -1$ implies $\lambda_2(p) = -1$, or, what is the same, that $\lambda_2(p) = 1$ implies $C_p(2) = 1$. That is, it is to be shown that $p \equiv 7 \bmod 8$ and $p \equiv 1 \bmod 8$ both imply $C_p(2) = 1$.

If $p \equiv 7 \bmod 8$, the signatures of the pivotal modules $[p, \sqrt{p}]$ and $[\frac{p-1}{2}, 1+\sqrt{p}]$ are both $--$ because $\lambda_1(p) = -1$ and $\lambda_1(\frac{p-1}{2}) = -1$; therefore, the signature of $[2, 1+\sqrt{p}]$, the one remaining pivotal module other than $[1]$, must be $++$, which means in particular that $C_p(2) = 1$.

Finally, if $p \equiv 1 \bmod 8$, then either $[8, 1+\sqrt{p}]$ or $[8, 5+\sqrt{p}]$ is primitive (if $p = 17$, then $[8, 5+\sqrt{p}]$ is primitive and for all greater primes congruent to 1 mod 8 the difference between $\frac{p-25}{8}$ and $\frac{p-1}{8}$ is 3, so exactly one of the two is odd). The signature of this primitive module is $+$ (because $A \equiv 1 \bmod 4$), which is to say that $C_p(8) = 1$; thus, $C_p(2)^3 = 1$, which implies $C_p(2) = 1$, as was to be shown. □

Exercise for Chapter 28

The law of quadratic reciprocity can be used to evaluate $C_p(A)$ for large values of p and A. For example, $C_{67}(102) = C_{67}(2 \cdot 3 \cdot 17) = C_{67}(2) \cdot C_{67}(3) \cdot C_{67}(17) = \lambda_2(67) \cdot (-C_3(67)) \cdot C_{17}(67) = \lambda_2(3) \cdot (-1) \cdot C_{17}(16) = (-1) \cdot (-1) \cdot (+1) = 1$, so 102 is a square mod 67. (Or, one could say $C_{67}(102) = C_{67}(35) = C_{67}(5) \cdot C_{67}(7) = C_5(67) \cdot (-C_7(67)) = C_5(2) \cdot (-C_7(4)) = (-1) \cdot (-1) = 1$.) Pose problems of this sort for yourself and solve them. (In addition to quadratic reciprocity you may want to use $C_p(x^2y) = C_p(y)$ whenever $x \not\equiv 0 \bmod p$.) Your answers can always be checked using Euler's criterion (Chapter 21).

Chapter 29

Proof of the Main Theorem

The Main Theorem (see Chapter 25) states that if A is squarefree and if p is an odd prime that does not divide A, then $C_p(A)$ is 1 if and only if the signature of p for A contains an even number of minuses. In formulas,

$$C_p(A) = C_{A_1}(p)C_{A_2}(p)\cdots C_{A_m}(p)\lambda(p), \tag{1}$$

where $A_1, A_2, \ldots, A_m$ are the odd prime factors of A and where the last factor $\lambda(p)$ is 1 when $A \equiv 1 \bmod 4$, $\lambda_1(p)$ when $A \equiv 3 \bmod 4$, $\lambda_2(p)$ when $A \equiv 2 \bmod 8$, and $\lambda_3(p)$ when $A \equiv 6 \bmod 8$. (See the end of Chapter 24 for the definitions of λ_1, λ_2, and λ_3.)

Proof. When $A = 2$, formula (1) is simply the proposition of the last chapter.

When A is an odd prime, formula (1) follows from quadratic reciprocity in the following way. When $A \equiv 1 \bmod 4$, formula (1) becomes $C_p(A) = C_A(p)$, which follows from quadratic reciprocity. When $A \equiv 3 \bmod 4$, it becomes $C_p(A) = C_A(p)\lambda_1(p)$; thus, it states $C_p(A) = C_A(p)$ when $p \equiv 1 \bmod 4$ and $C_p(A) = -C_A(p)$ when $p \equiv 3 \bmod 4$, both of which follow from quadratic reciprocity.

For any odd, squarefree A, say $A = A_1 A_2 \cdots A_m$ where A_1, A_2, $\ldots$, A_m are distinct odd primes, one therefore has

$$C_p(A) = C_p(A_1)C_p(A_2)\cdots C_p(A_m) = C_{A_1}(p)C_{A_2}(p)\cdots C_{A_m}(p)\lambda_1(p)^i$$

where i is the number of the primes $A_1, A_2, \ldots, A_m$ that are 3 mod 4. Since $\lambda_1(p)^2 = 1$, it follows that $C_p(A)$ is given by formula (1) where $\lambda(p) = 1$ when i is even and $\lambda(p) = \lambda_1(p)$ when i is odd. Since $A \equiv 1 \bmod 4$ when i is even and $A \equiv 3 \bmod 4$ when i is odd, this is the formula that was to be proved in the case in which A is odd.

Finally, if A is even, one finds similarly that

$$C_p(A) = C_p(A_1)C_p(A_2)\cdots C_p(A_m)C_p(2)$$

is given by (1) when the last factor is $C_p(2)\lambda_1(p)^i$ (when i is as before). Since this factor is $\lambda_2(p)$ when i is even and $\lambda_3(p)$ when i is odd, it is the last factor in the statement of the Main Theorem, and the proof is complete. □

As was shown in Chapter 25, Euler's law is a corollary of the Main Theorem. As was shown in Exercise 3 of Chapter 21, quadratic reciprocity—in the form Gauss stated it—is a corollary of Euler's law.

Chapter 30

The Theory of Binary Quadratic Forms

The modules of hypernumbers that are used to solve $A\square + B = \square$ and prove "Euler's law" in the preceding chapters provide a different approach to one of the classic topics of number theory, the theory of binary quadratic forms.

The theory of binary quadratic forms does not use hypernumbers, but it does use *integers*—not just the numbers 0, 1, 2, ... that have been used in this book so far but the negative integers -1, -2, -3, ... as well. In addition, it uses *computations with polynomials with integer coefficients*. (Chapter 14 made limited use of polynomials in *one* indeterminate with *number* coefficients, but now polynomials in *several* indeterminates with *integer* coefficients will be needed.) The reader will very likely be well acquainted with such computations, even though they have not been used so far in the book.

A **form** is a homogeneous polynomial—that is, a polynomial in which all terms have the same degree. When that degree is 2, the form is called a **quadratic form**. Finally, when a quadratic form involves just two indeterminates, it is called a **binary quadratic form**. Since only binary quadratic forms will be considered in this chapter, they will simply be called *forms*.

In short, a form is a polynomial $Ax^2 + Bxy + Cy^2$ in which A, B, and C are integers and x and y are indeterminates. It will be convenient to impose a few further conditions on the forms that are to be considered. First, the coefficient of the first term will be assumed to be *positive.* (This condition takes advantage of the fact that the properties of $Ax^2 + Bxy + Cy^2$ determine the properties of $-Ax^2 - Bxy - Cy^2$, so the sign can be chosen to make $A \geq 0$. For a reason that is about to be explained, the case $A = 0$ can be ignored.) Second, it will be assumed that the coefficient of the term that contains both indeterminates is *even.* (Again, the properties of $Ax^2 + Bxy + Cy^2$ determine the properties of $2Ax^2+2Bxy+2Cy^2$, so there is no loss of generality in assuming that the middle coefficient is an even integer. This choice agrees with Gauss's notation—he wrote forms as $ax^2 + 2bxy + cy^2$—and, as will be seen, it meshes well with the notation used for modules of hypernumbers in the previous chapters.) Third, it will be assumed that *the discriminants of the forms considered are not squares.* Here the **discriminant** of a form $ax^2 + 2bxy + cy^2$ is the integer $b^2 - ac$. (The rationale for this assumption is that forms whose discriminants are square are *decomposable* as products of two linear factors with rational coefficients and are therefore in some sense not truly quadratic forms.[1] Note that the assumption $b^2 - ac \neq \square$ implies $a \neq 0$.) Finally, it will be assumed that the *discriminant is positive,* simply because these are the only forms that can be treated using the hypernumbers that have been considered in the previous chapters. (As will be seen in the exercises, hypernumbers that involve $\sqrt{A}$ for negative A can be treated in exactly the same way as those for positive A, and the problem of determining whether two modules of such hypernumbers are equivalent—as well as the solution of $A\square + B = \square$—is in fact *easier* when $A < 0$. Once these facts are established, the theory of forms with negative discriminants proceeds in the same way as the theory for positive discriminants.)

---

[1] Characteristically, Gauss did *not* exclude square discriminants; instead he dealt with this case separately. He called $b^2 - ac$ the *determinant* instead of the discriminant of the form, a rare case in which Gauss's terminology has not been adopted by later generations. If $b^2 - ac = k^2$ for some integer k, the formula $a(ax^2 + 2bxy + cy^2) = (ax + by)^2 - k^2y^2 = (ax + by + ky)(ax + by - ky)$ shows that $ax^2 + 2bxy + cy^2$ has the decomposition $(ax + by + ky)(ax + by - ky)/a$.

Gauss defined the concept of **equivalence** of forms in the following way. Substitution of

(1)
$$u = qx + ry$$
$$v = sx + ty$$

where q, r, s, and t are integers, in a form $\alpha u^2 + 2\beta uv + \gamma v^2$, where α, β, and γ are integers, gives a form $ax^2 + 2bxy + cy^2$ in which a, b, and c are integers. Specifically, the values of a, b, and c are given by the matrix formula

(2)
$$\begin{bmatrix} a & b \\ b & c \end{bmatrix} = \begin{bmatrix} q & s \\ r & t \end{bmatrix} \begin{bmatrix} \alpha & \beta \\ \beta & \gamma \end{bmatrix} \begin{bmatrix} q & r \\ s & t \end{bmatrix}.$$

The matrix of coefficients $\begin{bmatrix} q & r \\ s & t \end{bmatrix}$ is *invertible*, with an inverse that has integer entries, if and only if its determinant $qt - rs$ is ± 1. When this is the case, its inverse transforms $ax^2 + 2bxy + cy^2$ back to $\alpha u^2 + 2\beta uv + \gamma v^2$. Prior to Gauss, two forms were considered to be "the same" if such an invertible way of transforming one of them into the other was possible. Gauss was the first to realize that this seemingly natural definition of the equivalence of two forms led to confusions that were avoided if one required that two forms be considered **equivalent** *only* if they could be transformed into one another by a transformation (1) in which the determinant $qt - rs$ of the matrix of coefficients was 1.

From the point of view of *forms* Gauss's definition of equivalence is undeniably artificial—it depends on the order of the variables because it requires distinguishing the change of variables $x = u$, $y = v$ from the change of variables $x = v$, $y = u$—but from the point of view of modules of hypernumbers the theorem below shows that the condition $qt - rs = 1$ is entirely natural.

Problem. *Given two forms, determine whether they are equivalent and, if so, find all possible transformations of one into the other by matrices with determinant* 1.

Since the determinant of a product is the product of the determinants, equation (2) shows that equivalent forms must have the same discriminant $b^2 - ac = \beta^2 - \alpha\gamma$ (because the determinant on the right is $ac - b^2$ and the determinant on the left is $\alpha\gamma - \beta^2$). Thus, the problem is to determine whether two forms *with the same discriminant* are equivalent.

Theorem. *Let $ax^2 + 2bxy + cy^2$ and $\alpha u^2 + 2\beta uv + \gamma v^2$ be given forms with the same discriminant, say $A = b^2 - ac = \beta^2 - \alpha\gamma$. They are equivalent if and only if the modules of hypernumbers*[2] $[a, b + \sqrt{A}]$ *and* $[\alpha, \beta + \sqrt{A}]$ *are equivalent.*

Since the method of Chapter 23 enables one to determine whether two modules are equivalent, this theorem solves the problem of determining whether two given forms are equivalent. Moreover, when the forms are equivalent, the method of the proof gives an algorithm for finding all changes of variables (1) with determinant 1 that transform one into the other, or, more precisely, reduces the problem of finding all such changes of variables to the problem of finding all hypernumbers $Y + X\sqrt{A}$ which satisfy $[Y + X\sqrt{A}][a, b + \sqrt{A}] = [a][\alpha, \beta + \sqrt{A}]$; this problem is solved in Chapter 23 in the special case in which $[\alpha, \beta + \sqrt{A}]$ is stable, and the general case follows easily from this case.

Proof. Assume first that $[a, b + \sqrt{A}]$ and $[\alpha, \beta + \sqrt{A}]$ are equivalent. As was shown in Chapter 23, there is then an equivalence of the particular form $[Y + X\sqrt{A}][a, b + \sqrt{A}] = [a][\alpha, \beta + \sqrt{A}]$, where X and Y are numbers for which $Y^2 > AX^2$. Given such a hypernumber $Y + X\sqrt{A}$, the matrix defined by the formula

$$(3) \qquad \begin{bmatrix} q & r \\ s & t \end{bmatrix} = \begin{bmatrix} \frac{Y - \beta X}{\alpha} & \frac{Yb + AX - \beta Y - \beta bX}{a\alpha} \\ X & \frac{Y + bX}{a} \end{bmatrix}$$

has integer entries, has determinant 1, and satisfies equation (2), as can be verified in the following way.

---

[2]When b is negative, $b + \sqrt{b^2 - ac}$ is not a hypernumber. This problem can be overcome either by allowing hypernumbers to have negative coefficients or by replacing $ax^2 + 2bxy + cy^2$ with an equivalent form in which a is unchanged and b is positive, which is easy to do—see Exercise 1. Since the module $[a, b + \sqrt{A}]$ depends only on the value of $b \bmod a$, the meaning of $[a, b + \sqrt{A}]$ for negative b is clear.

To say that $a(Y + X\sqrt{A})$ is 0 mod $[a][\alpha, \beta + \sqrt{A}]$ is to say that $aX \equiv 0 \bmod a$ and $aY \equiv \beta aX \bmod a\alpha$ (by the proposition of Chapter 18), which is simply to say $Y \equiv \beta X \bmod \alpha$; in short, q in formula (3) is an integer. To say that $(Y + X\sqrt{A})(b + \sqrt{A}) = Yb + XA + (Y + Xb)\sqrt{A}$ is 0 mod $[a][\alpha, \beta + \sqrt{A}]$ is to say that $Y + Xb \equiv 0 \bmod a$ and $Yb + XA \equiv \beta(Y + Xb) \bmod a\alpha$, or, in short, that t and r are integers. Not only is s an integer, it is a number. Since $qt - rs = \frac{1}{a\alpha}\big((Y - \beta X)(Y + bX) - X(Yb + AX - \beta Y - \beta bX)\big) = \frac{1}{a\alpha}(Y^2 - AX^2)$, the statement that the determinant of $\begin{bmatrix} q & r \\ s & t \end{bmatrix}$ is 1 is the statement that $Y^2 - AX^2 = a\alpha$, which follows from the fact that the norm[3] of $[Y + X\sqrt{A}][a, b + \sqrt{A}] = [a][\alpha, \beta + \sqrt{A}]$ is $[(Y^2 - AX^2)a][a, 2b, c, b + \sqrt{A}] = [a^2\alpha][\alpha, 2\beta, \gamma, \beta + \sqrt{A}]$.

Finally, the needed equation (2) follows from noting that the substitution (1) carries $a\big(\alpha u + (\beta + \sqrt{A})v\big)$ to $a\alpha(qx + ry) + a\beta(sx + ty) + a(sx + ty)\sqrt{A} = a(Y - \beta X)x + (Yb + AX - \beta Y - \beta bX)y + a\beta Xx + \beta(Y + bX)y + \big(aXx + (Y + bX)y\big)\sqrt{A} = Y(ax + by + y\sqrt{A}) + X(-a\beta x + Ay - \beta by + a\beta x + \beta by + ax\sqrt{A} + by\sqrt{A}) = Y(ax + by + y\sqrt{A}) + X(Ay + ax\sqrt{A} + by\sqrt{A}) = (Y + X\sqrt{A})(ax + by + y\sqrt{A})$. (This fact—that substitution of (1) in $a\big(\alpha u + (\beta + \sqrt{A})v\big)$ gives this result—is the source of formula (3).) Therefore, the same substitution in $a\big(\alpha u + (\beta - \sqrt{A})v\big)$ carries it to $(Y - X\sqrt{A})(ax + by - y\sqrt{A})$, so it also carries $a^2\big((\alpha u + \beta v)^2 - Av^2\big)$ to $(Y^2 - AX^2)\big((ax + by)^2 - Ay^2\big)$; division by $a^2\alpha = a(Y^2 - AX^2)$ then gives the required identity $\alpha u^2 + 2\beta uv + \gamma v^2 = ax^2 + 2bxy + cy^2$.

Conversely, suppose $[a, b + \sqrt{A}]$ and $[\alpha, \beta + \sqrt{A}]$ are given and suppose a 2×2 matrix of integers $\begin{bmatrix} q & r \\ s & t \end{bmatrix}$ with determinant 1 is given for which (2) holds (where, of course, c and γ are defined by $b^2 - ac = A$ and $\beta^2 - \alpha\gamma = A$, respectively). The formulas $X = s$ and $Y = at - bs$ determine *integer* values of X and Y for which $[Y + X\sqrt{A}][a, b + \sqrt{A}] = [a][\alpha, \beta + \sqrt{A}]$ in the following way.

[3]See Chapter 20 for the definition of the norm of a module.

Equation (2) implies, because the determinant of the matrix of coefficients is 1, that

$$\begin{bmatrix} t & -s \\ -r & q \end{bmatrix}\begin{bmatrix} a & b \\ b & c \end{bmatrix}\begin{bmatrix} t & -r \\ -s & q \end{bmatrix} = \begin{bmatrix} \alpha & \beta \\ \beta & \gamma \end{bmatrix},$$

which implies, when the entries of the first row in the product on the left are computed, that $\alpha = at^2 - 2bst + cs^2$ and $\beta = -art + bqt + brs - cqs$. On the other hand, $[Y + X\sqrt{A}][a, b + \sqrt{A}] = [at - bs + s\sqrt{A}][a, b + \sqrt{A}] = [a][at - bs + s\sqrt{A}, \frac{abt-b^2s+As}{a} + \frac{at-bs+bs}{a} \cdot \sqrt{A}] = [a][at - bs + s\sqrt{A}, bt - cs + t\sqrt{A}]$. Let $\mathcal{A} = at - bs + s\sqrt{A}$ and $\mathcal{B} = bt - cs + t\sqrt{A}$ so that $[Y + X\sqrt{A}][a, b + \sqrt{A}] = [a][\mathcal{A}, \mathcal{B}]$. Simple computation gives $t\mathcal{A} - s\mathcal{B} = at^2 - bst - bst + cs^2 = \alpha$ and $-r\mathcal{A} + q\mathcal{B} = -art + brs + bqt - cqs + (-rs + qt)\sqrt{A} = \beta + \sqrt{A}$, so $[\mathcal{A}, \mathcal{B}] = [\mathcal{A}, \mathcal{B}, \alpha, \beta + \sqrt{A}]$. Since $\begin{bmatrix} q & s \\ r & t \end{bmatrix}$ is the inverse of $\begin{bmatrix} t & -s \\ -r & q \end{bmatrix}$, the equations $t\mathcal{A} - s\mathcal{B} = \alpha$ and $-r\mathcal{A} + q\mathcal{B} = \beta + \sqrt{A}$ imply $\mathcal{A} = q\alpha + s(\beta + \sqrt{A})$ and $\mathcal{B} = r\alpha + t(\beta + \sqrt{A})$, which means that $\mathcal{A}$ and $\mathcal{B}$ can be dropped to find $[\mathcal{A}, \mathcal{B}] = [\alpha, \beta + \sqrt{A}]$, as was to be shown.

If X and Y are both positive, or if $X = 0$, this equation $[Y + X\sqrt{A}][a, b + \sqrt{A}] = [a][\alpha, \beta + \sqrt{A}]$ proves that $[a, b + \sqrt{A}] \sim [\alpha, \beta + \sqrt{A}]$. If both X and Y are negative, the equivalent equation $[-Y - X\sqrt{A}][a, g + \sqrt{A}] = [a][\alpha, \beta + \sqrt{A}]$ yields the same conclusion. Finally, if X and Y have opposite signs, then multiplication of both sides by $[Y - X\sqrt{A}]$ gives $[Y^2 - AX^2][a, b + \sqrt{A}] = [a][Y - X\sqrt{A}][\alpha, \beta + \sqrt{A}]$; since $Y^2 - AX^2 = a\alpha$ (take norms on both sides), this equation implies $[\alpha][a, b + \sqrt{A}] = [Y - X\sqrt{A}][\alpha, \beta + \sqrt{A}]$. Since $-X$ and Y have the same sign, the desired conclusion $[a, b + \sqrt{A}] \sim [\alpha, \beta + \sqrt{A}]$ follows from the cases already proved when the roles of $[a, b + \sqrt{A}]$ and $[\alpha, \beta + \sqrt{A}]$ are reversed. □

Exercises for Chapter 30

1. Show that every form $ax^2 + 2bxy + cy^2$ in which a is positive is equivalent to one in which a is unchanged and b is positive.

2. Gauss's first example [**G**, Art. 158] of a pair of equivalent forms is $2x^2 - 8xy + 3y^2$ and $-13u^2 - 12uv - 2v^2$, for which he gives

the explicit equivalence $u = 2x - y$, $v = -3x + 2y$. The method of this chapter does not apply directly to this example because $-13u^2 - 12uv - 2v^2$ does not correspond to a module. The form $2x^2 - 8xy + 3y^2$ corresponds to $[2, -4 + \sqrt{10}] = [2, \sqrt{10}]$, as does the form $2x^2 - 5y^2$, so these two forms are equivalent. The substitution $X = y$, $Y = -x$ given by the matrix $\begin{bmatrix} 0 & 1 \\ -1 & 0 \end{bmatrix}$ with determinant 1 shows that $-5X^2 + 2Y^2$ and $2x^2 - 5y^2$ are equivalent. Therefore, Gauss's equivalence can be found by using the method of the chapter to find an equivalence of $5X^2 - 2Y^2$ and $13u^2 + 12uv + 2v^2$. Carry out the explicit construction of an equivalence these steps indicate. (There is no reason to expect, *a priori*, that the equivalence found in this way will be the one Gauss gives.)

3. For what primes p is the form $x^2 - py^2$ equivalent in Gauss's sense to the form $-u^2 + pv^2$?

Exercises on Hypernumbers with Negative A

In order to deal with hypernumbers with negative A, it makes sense to start with *integers* instead of numbers in the sense of Chapter 1. Such hypernumbers will be written $y + x\sqrt{-A}$, where x and y are integers and $-A$ is a negative integer, and to make the distinction clear, they will be called **hyperintegers**. The following exercises show that the theory of modules of hyperintegers can be developed in ways that are parallel to the methods used in the previous chapters.

Hyperintegers can be added and multiplied in the obvious ways. Moreover, they can be *subtracted*, so that $m \equiv n \bmod [a_1, a_2, \dots, a_k]$ can be defined either in the way it is defined for hypernumbers in Chapter 17 or defined more simply as meaning that $m - n = r_1a_2 + r_2a_2 + \cdots + r_ka_k$ where $r_1, r_2, \dots, r_k$ are hyperintegers. As in Chapter 17, $[a_1, a_2, \dots, a_k] = [b_1, b_2, \dots, b_l]$ means that two hyperintegers are congruent mod $[a_1, a_2, \dots, a_k]$ if and only if they are congruent mod $[b_1, b_2, \dots, b_l]$.

4. Prove that any module $[a_1, a_2, \dots, a_k]$ that is not equal to $[0]$ (that is, the corresponding congruence relation is not equality) is equal to one and only one module of the form $[e][f, g + \sqrt{-A}]$, where

e and f are positive integers, $0 \le g < f$, and $g^2 \equiv -A \bmod f$. Such a module of hyperintegers will be said to be in **canonical form**.

5. Let a module of hyperintegers be called **principal** if it can be written in the form $[y + x\sqrt{-A}]$ (so there is no condition on the sizes or signs of x and y), and let two modules be called **equivalent** if they can be made equal by multiplying them by principal modules. State and prove an analog of the comparison algorithm for modules of hyperintegers.

6. Prove that if p is a prime and $p \equiv 1 \bmod 4$, then $[p, g+\sqrt{-1}]$ is a module in canonical form for some g. Show that repeated application of the comparison algorithm to $[p, g + \sqrt{-1}]$ reaches the module $[1]$ (after which it merely repeats $[1]$ endlessly) and that the formula of the theorem of Chapter 19 then gives a representation of p as a sum of two squares. This is the classic fact of number theory, first proved by Euler, that *a prime that is* 1 mod 4 *can be written as a sum of two squares.*

7. Use a method like the one of the previous exercise to prove that *if p is prime and $p \equiv 1$ or* 3 mod 8, *then p can be written as a square plus twice a square.*

8. Similarly, if p is 1 mod 3, then $p = \square + 3\square$.

9. By Euler's law, the value of $C_p(5)$ depends only on the value of p mod 20. One easily finds $C_p(5) = 1$ if and only if $p \equiv 1$, 3, 7, or 9 mod 20. In this case it is not always true, however, that $p = \square + 5\square$. Try several cases and discover what is true. (Hint: The class group has two elements.)

Chapter 31

Composition of Binary Quadratic Forms

The theorem of Chapter 30 shows that Gauss's partition of forms into equivalence classes coincides with the partition of modules of hypernumbers into their equivalence classes. But between these parallel theories—the "module" theory and the "form" theory—the module theory has a distinct advantage: Modules can be *multiplied* but forms cannot.

Disquisitiones Arithmeticae develops the theory strictly in terms of forms, which means that this operation—so easily described in terms of modules—must be described as an operation with forms. This is the role of *composition of forms* in Gauss's theory.

Since the "module" point of view has been used in the earlier chapters to give complete derivations of the solution of $A\square + B = \square$, of "Euler's law," and of the law of quadratic reciprocity, it is not necessary here, as it was for Gauss, to give a full theory of composition of forms in connection with these derivations; a brief explanation of an algorithm for composing forms will suffice.

When Brahmagupta's formula is stated

$$\begin{aligned}&\text{if } X = xv + yu \text{ and } Y = yv + Axu,\\ &\quad\text{then } Y^2 - AX^2 = (y^2 - Ax^2)(v^2 - Au^2)\end{aligned}$$

(here A is to be thought of as an integer—preferably a positive integer not a square—but in fact the formula is true when A, as well as x, y, u, and v, is regarded as an indeterminate), it is a special case of the composition of forms.

More generally, a **composition** of two forms $ax^2+2bxy+cy^2$ and $\alpha u^2+2\beta uv+\gamma v^2$ is a way of defining polynomials X and Y in x, y, u, and v with the property that $(ax^2+2bxy+cy^2)(\alpha u^2+2\beta uv+\gamma v^2)$ can be expressed as a form $\mathcal{A}X^2+2\mathcal{B}XY+\mathcal{C}Y^2$ in X and Y. Here a, b, c, α, β, γ, $\mathcal{A}$, $\mathcal{B}$, and $\mathcal{C}$ are integers, x, y, u, and v are indeterminates, and X and Y are polynomials in the indeterminates with integer coefficients. General considerations dictate that X and Y must be bilinear in (x, y) and (u, v), which is to say that

$$X = p_0xu + p_1xv + p_2yu + p_3yv$$
$$Y = q_0xu + q_1xv + q_2yu + q_3yv$$

where the p's and q's are integers. A composition of $ax^2+2bxy+cy^2$ and $\alpha u^2+2\beta uv+\gamma v^2$ consists of polynomials X and Y of this form and integers $\mathcal{A}$, $\mathcal{B}$, and $\mathcal{C}$, for which the polynomial identity

$$(1)\quad \mathcal{A}X^2+2\mathcal{B}XY+\mathcal{C}Y^2 = (ax^2+2bxy+y^2)(\alpha u^2+2\beta uv+\gamma v^2)$$

holds.[1]

Gauss proved (under very mild assumptions of nondegeneracy) that the existence of such a formula implies that *the ratio of the discriminants* b^2-ac *and* $\beta^2-\alpha\gamma$ *of the two given forms must be a ratio of squares,* which is to say that there must be nonzero numbers s and σ for which $s^2(b^2-ac)=\sigma^2(\beta^2-\alpha\gamma)$. Conversely, when this condition is met, one can *explicitly* construct a composition formula (1) in the following way.

First, a few simplifying assumptions. Since a composition formula for $ax^2+2bxy+cy^2$ and $\alpha u^2+2\beta uv+\gamma v^2$ implies one for $-ax^2-$

[1]Gauss also stipulates that the six 2×2 minors of the 2×4 matrix of p's and q's must be relatively prime and the first two of them must be positive. The method explained below always yields formulas that meet these technical conditions, which are needed to avoid certain degenerate cases and to make the operation of composition consistent with the stronger meaning of equivalence that Gauss had introduced.

$2bxy - cy^2$ and $\alpha u^2 + 2\beta uv + \gamma v^2$ (use the same X and Y and simply reverse the signs of $\mathcal{A}$, $\mathcal{B}$, and $\mathcal{C}$), there is no loss of generality in assuming that $a \geq 0$ and $\alpha \geq 0$. As before, it will be assumed that the discriminants are not squares (even though Gauss considered that case as well), so, in particular, neither a nor α is zero. Since a composition formula for $ax^2+2bxy+cy^2$ and $\alpha u^2+2\beta uv+\gamma v^2$ follows easily from a composition formula for $a(x+ny)^2+2b(x+ny)y+cy^2 = ax^2 + 2(b + na)xy + (c + 2bn + an^2)y^2$ and $\alpha u^2 + 2\beta uv + \gamma v^2$, one can also assume without loss of generality that b (and even c) is positive. Finally, it will be assumed that the common value A of $s^2(b^2 - ac) = \sigma^2(\beta^2 - \alpha\gamma)$ is positive so that the theory of modules of hypernumbers developed in the earlier chapters applies. (The case $A < 0$ requires only very minor adjustments.)

Given $ax^2 + 2bxy + cy^2$ and $\alpha u^2 + 2\beta uv + \gamma v^2$ satisfying these conditions, compute the product of modules

$$[sa, sb + \sqrt{A}][\sigma\alpha, \sigma\beta + \sqrt{A}] = [E][F, G + \sqrt{A}] \tag{2}$$

where the right side is in canonical form.

Theorem. *Under these conditions, the formula*

$$(sax + sby + y\sqrt{A})(\sigma\alpha u + \sigma\beta v + v\sqrt{A}) = E(FX + GY + Y\sqrt{A}) \tag{3}$$

determines polynomials X and Y in x, y, u, and v with the property that $(ax^2+2bxy+cy^2)(\alpha u^2+2\beta uv+\gamma v^2)$ can be expressed as a form in X and Y. In fact, explicit formulas for integers $\mathcal{A}$, $\mathcal{B}$, and $\mathcal{C}$ will be given that satisfy (1).

Proof. The definitions of E, F, and G in (2) imply that $sa \cdot \sigma\alpha$, $sa(\sigma\beta+\sqrt{A})$, $\sigma\alpha(sb+\sqrt{A})$, and $(sb+\sqrt{A})(\sigma\beta+\sqrt{A})$ are all divisible by $[E][F, G+\sqrt{A}]$. In other words, sa, $\sigma\alpha$, and $sb+\sigma\beta$ are all divisible by E, and the congruences $sa \cdot \sigma\alpha \equiv 0 \bmod EF$, $sa \cdot \sigma\beta \equiv Gsa \bmod EF$, $\sigma\alpha \cdot sb \equiv G\sigma\alpha \bmod EF$, and $sb \cdot \sigma\beta + A \equiv G(sb + \sigma\beta) \bmod EF$ all hold.

When one equates the coefficients of $\sqrt{A}$ on the two sides of (3), one finds that Y must satisfy

$$(sax + sby)v + (\sigma\alpha u + \sigma\beta v)y = EY,$$

which shows that

$$Y = \frac{sa}{E} \cdot xv + \frac{\sigma\alpha}{E} \cdot yu + \frac{sb + \sigma\beta}{E} \cdot yv,$$

a formula which determines Y as a polynomial in x, y, u, and v with integer coefficients.

The terms of (3) without $\sqrt{A}$ give

$$(sax + sby)(\sigma\alpha u + \sigma\beta v) + Ayv = EFX + EGY.$$

Subtract $EGY = G\big(sa \cdot xv + \sigma\alpha \cdot yu + (sb + \sigma\beta)yv\big)$ from both sides and divide by EF to find

$$\frac{sa \cdot \sigma\alpha}{EF} \cdot xu + \frac{sa(\sigma\beta - G)}{EF} \cdot xv + \frac{\sigma\alpha(sb - G)}{EF} \cdot yu + \frac{sb \cdot \sigma\beta + A - G(sb + \sigma\beta)}{EF} \cdot yv = X,$$

which expresses X as a polynomial in x, y, u, and v with integer coefficients.

The polynomials X and Y defined in this way also satisfy $(sax + sby - y\sqrt{A})(\sigma\alpha u + \sigma\beta v - v\sqrt{A}) = E(FX + GY - Y\sqrt{A})$ and when this equation is multiplied by (3), the result is

$$\big((sax+sby)^2 - Ay^2\big)\big((\sigma\alpha u+\sigma\beta v)^2 - Av^2\big) = E^2\big((FX+GY)^2 - AY^2\big).$$

Since $A = s^2(b^2 - ac) = \sigma^2(\beta^2 - \alpha\gamma)$, division of this equation by $s^2\sigma^2 a\alpha$ gives

$$(ax^2 + 2bxy + cy^2)(\alpha u^2 + 2\beta uv + \gamma v^2) = \frac{E^2F^2X^2 + 2E^2FGXY + E^2(G^2 - A)Y^2}{s^2\sigma^2 a\alpha},$$

which is a composition formula because the three coefficients in the numerator on the right are divisible by the denominator so the right side has the form $\mathcal{A}X^2 + 2\mathcal{B}XY + \mathcal{C}Y^2$, as can be proved in the following way.

The norm of equation (2) is

$$(4) \qquad [sa \cdot \sigma\alpha][sm, sb + \sqrt{A}][\sigma\mu, \sigma\beta + \sqrt{A}] = [E^2F][M, G + \sqrt{A}],$$

where m, μ, and M are defined by $[m] = [sa, 2sb, sc]$, $[\mu] = [\sigma\alpha, 2\sigma\beta, \sigma\gamma]$, and $[M] = [F, 2G, \frac{|G^2-A|}{F}]$. The norm of (4), in turn, is
(5)
$$[sa\cdot\sigma\alpha]^2[sm\cdot\sigma\mu][sm, sb+\sqrt{A}][\sigma\mu, \sigma\beta+\sqrt{A}] = [E^2F]^2[M][M, G+\sqrt{A}]$$

because $2b \equiv 0 \bmod m$ and $b^2 \equiv A \bmod a\cdot c$, which implies $b^2 \equiv A \bmod m^2$, and similarly for μ and M. When (4) is used to put the left side of (5) in canonical form, one finds

$$[sa\cdot\sigma\alpha][sm\cdot\sigma\mu][E^2F][M, G+\sqrt{A}] = [E^2F]^2[M][M, G+\sqrt{A}]$$

which amounts simply to the statement that $sa\cdot\sigma\alpha\cdot sm\cdot\sigma\mu = E^2FM$. In view of the definition of M, the desired conclusion that $s^2\sigma^2a\alpha$ divides E^2F^2, $2E^2FG$, and $E^2(G^2 - A)$ follows. □

One composition formula for $ax^2+2bxy+cy^2$ and $\alpha u^2+2\beta uv+\gamma v^2$ implies infinitely many others, because the substitution

$$U = qX + rY$$
$$V = sX + tY$$

where q, r, s, and t are integers satisfying $qt - rs = 1$ gives another composition formula when $X = tU - rV$ and $Y = -sU + qV$ are substituted in the known composition formula. Gauss proved, again under very mild nondegeneracy assumptions, that *one* composition formula implies *all others* in this way. Therefore, the above construction determines all composition formulas for $ax^2 + 2bxy + cy^2$ and $\alpha u^2 + 2\beta uv + \gamma v^2$. In short, *any two forms* $\mathcal{A}X^2 + 2\mathcal{B}XY + \mathcal{C}Y^2$ *that compose* $ax^2 + 2bxy + cy^2$ *and* $\alpha u^2 + 2\beta uv + \gamma v^2$ *are equivalent.* Therefore, *all* compositions of two given forms can be found once *one* has been found by the construction above.

The proof of this key theorem is only the beginning of the problems associated with doing the entire theory in terms of forms. A much greater obstacle is the proof that composition is *associative* in the appropriate sense. But if one chooses to regard compositions of forms as a cumbersome way to accomplish what is accomplished more easily and naturally using computations with modules of hypernumbers—for which the associativity of multiplication is self-evident—one can ignore these difficulties.

Exercises for Chapter 31

1. Find a general formula for composing $x^2 - Ay^2$ and $\alpha u^2 + 2\beta uv + \gamma v^2$ when $\beta^2 - \alpha\gamma = A$.

2. Find a formula for the composition of $3x^2 + 2xy + 47y^2$ and $9u^2+14uv+21v^2$. (See [**G,** Art. 243]. For these forms $b^2-ac = -140$, so computations with hypernumbers $y + x\sqrt{-140}$ are needed. It is easy to guess what to do, and the result is an explicit formula that can be verified without reference to hypernumbers.)

3. Almost all treatments of the composition of forms since Gauss's original treatment have ignored the composition of *forms* and have merely composed *equivalence classes* of forms. For Gauss, who had done the hard work of defining the composition of forms in full, the definition of the composite of two equivalence classes followed simply from:

Theorem. *The equivalence class of the composite of two given forms depends only on the equivalence classes of the given forms.*

Otherwise stated:

Theorem. *If there is a composition formula in which $\mathcal{A}X^2+2\mathcal{B}XY+\mathcal{C}Y^2$ is a composite of $ax^2 + 2bxy + cy^2$ and $\alpha u^2 + 2\beta uv + \gamma v^2$ and another in which $\mathcal{A}_1X_1^2 + 2\mathcal{B}_1X_1Y_1 + \mathcal{C}_1Y_1^2$ is a composite of $a_1x_1^2 + 2b_1x_1y_1+cy_1^2$ and $\alpha u^2+2\beta uv+\gamma v^2$ and if $ax^2+2bxy+cy^2$ is equivalent to $a_1x_1^2 + 2b_1x_1y_1 + c_1y_1^2$, then the form $\mathcal{A}X^2 + 2\mathcal{B}XY + \mathcal{C}Y^2$ is equivalent to $\mathcal{A}_1X^2 + 2\mathcal{B}_1XY + \mathcal{C}_1Y^2$.*

Prove this theorem. (You will need to make use of the theorem of Gauss cited in the text to the effect that any two forms $\mathcal{A}X^2 + 2\mathcal{B}XY + \mathcal{C}Y^2$ that compose the same forms $ax^2 + 2bxy + cy^2$ and $\alpha u^2 + 2\beta uv + \gamma v^2$ are equivalent.)

Appendix

Cycles of Stable Modules

The tables on the following pages describe the stable modules for all values of A, except squares, up to $A = 111$. The first line gives the value of A, followed by the number of stable modules for that A that correspond to each number k for which $k^2 < A$ (see Chapter 22). The following line or lines give one module from each cycle of stable modules and the number of modules in the cycle. (A programmable calculator can easily generate all modules in a cycle when a single module is given.) For squarefree values of A, the signature of each cycle is also given. (See Chapter 24.)

For example, when $A = 33$, there are 24 stable modules, partitioned into 4 cycles. There are 15 pairs (k, f), namely, $(0,1)$, $(0,3)$, $(0,11)$, $(0,33)$, $(1,2)$, $(1,4)$, $(1,8)$, $(1,16)$, $(1,32)$, $(2,29)$, $(3,6)$, $(3,8)$, $(3,12)$, $(3,24)$, $(4,17)$, 6 of which satisfy either $k = 0$ or $2k = f$. The principal cycle, the cycle of $[1]$, contains 2 modules; there is one other primitive cycle, the cycle of $[33, \sqrt{33}]$, containing 12 modules, and two cycles that are not primitive, one containing 4 modules and the other containing 6 modules. Since 33 is squarefree, the signatures of all four cycles are given.

$\mathbf{A} = 2.$ $\quad 0\ (2)$

$[1, \sqrt{2}] + (2)$

$\mathbf{A} = 3.$ $\quad 0\ (2),\ 1\ (1)$

$[1, \sqrt{3}] ++ (1),\ [2, 1 + \sqrt{3}] -- (2)$

$\mathbf{A} = 5.$ $\quad 0\ (2),\ 1\ (3)$

$[1, \sqrt{5}] + (4),\ [2, 1 + \sqrt{5}]^* - (1)$

$\mathbf{A} = 6.$ $\quad 0\ (4),\ 1\ (2)$

$[1, \sqrt{6}] ++ (2),\ [2, \sqrt{6}] -- (4)$

$\mathbf{A} = 7.$ $\quad 0\ (2),\ 1\ (5)$

$[1, \sqrt{7}] ++ (2),\ [7, \sqrt{7}] -- (5)$

$\mathbf{A} = 8.$ $\quad 0\ (4),\ 1\ (2),\ 2\ (1)$

$[1, \sqrt{8}]\ (1),\ [8, \sqrt{8}]\ (4),\ [2, \sqrt{8}]^*\ (2)$

$\mathbf{A} = 10.$ $\quad 0\ (4),\ 1\ (4),\ 2\ (2)$

$[1, \sqrt{10}] ++ (6),\ [2, \sqrt{10}] -- (4)$

$\mathbf{A} = 11.$ $\quad 0\ (2),\ 1\ (5),\ 2\ (2)$

$[1, \sqrt{11}] ++ (3),\ [11, \sqrt{11}] -- (6)$

$\mathbf{A} = 12.$ $\quad 0\ (6),\ 1\ (2),\ 2\ (3)$

$[1, \sqrt{12}]\ (2),\ [3, \sqrt{12}]\ (6),\ [2, \sqrt{12}]^*\ (1),\ [6, \sqrt{12}]^*\ (2)$

$\mathbf{A} = 13.$ $\quad 0\ (2),\ 1\ (9),\ 2\ (2)$

$[1, \sqrt{13}] + (10),\ [2, 1 + \sqrt{11}]^* - (3)$

$\mathbf{A} = 14.$ $\quad 0\ (4),\ 1\ (2),\ 2\ (4)$

$[1, \sqrt{14}] ++ (2),\ [14, \sqrt{14}] -- (8)$

$\mathbf{A} = 15.$ $\quad 0\ (4),\ 1\ (5),\ 2\ (2),\ 3\ (1)$

$[1, \sqrt{15}] +++ (1),\ [15, \sqrt{15}] -+- (6),\ [2, 1 + \sqrt{15}] --+ (2),$
$[3, \sqrt{15}] +-- (3)$

$\mathbf{A} = 17.$ $\quad 0\ (2),\ 1\ (7),\ 2\ (2),\ 3\ (2)$

$[1, \sqrt{17}] + (8),\ [2, 1 + \sqrt{17}]^* - (5)$

$\mathbf{A} = 18.$ $\quad 0\ (6),\ 1\ (2),\ 2\ (4),\ 3\ (2)$

$[1, \sqrt{18}]\ (4),\ [18, \sqrt{18}]\ (8),\ [3, \sqrt{18}]^*\ (2)$

$\mathbf{A} = 19.$ $\quad 0\ (2),\ 1\ (9),\ 2\ (4),\ 3\ (2)$

$[1, \sqrt{19}] ++ (7),\ [19, \sqrt{19}] -- (10)$

$\mathbf{A} = 20.$ 0 (6), 1 (2), 2 (5), 3 (2)
$[1, \sqrt{20}]$ (2), $[20, \sqrt{20}]$ (8), $[2, \sqrt{20}]^*$ (4), $[4, 2+\sqrt{20}]^*$ (1)

$\mathbf{A} = 21.$ 0 (4), 1 (9), 2 (2), 3 (3)
$[1, \sqrt{21}]$ $++$ (4), $[21, \sqrt{21}]$ $--$ (10), $[2, 1+\sqrt{21}]^*$ $-+$ (1),
$[6, 3+\sqrt{21}]^*$ $+-$ (3)

$\mathbf{A} = 22.$ 0 (4), 1 (6), 2 (6), 3 (2)
$[1, \sqrt{22}]$ $++$ (6), $[22, \sqrt{22}]$ $--$ (12)

$\mathbf{A} = 23.$ 0 (2), 1 (5), 2 (2), 3 (4)
$[1, \sqrt{23}]$ $++$ (2), $[23, \sqrt{23}]$ $--$ (11)

$\mathbf{A} = 24.$ 0 (8), 1 (2), 2 (7), 3 (2), 4 (1)
$[1, \sqrt{24}]$ (1), $[24, \sqrt{24}]$ (8), $[8, \sqrt{24}]$ (3), $[4, 2+\sqrt{24}]$ (2),
$[2, \sqrt{24}]^*$ (2), $[4, \sqrt{24}]^*$ (4)

$\mathbf{A} = 26.$ 0 (4), 1 (4), 2 (4), 3 (2), 4 (2)
$[1, \sqrt{26}]$ $++$ (10), $[2, \sqrt{26}]$ $--$ (6)

$\mathbf{A} = 27.$ 0 (4), 1 (5), 2 (2), 3 (5), 4 (2)
$[1, \sqrt{27}]$ (5), $[27, \sqrt{27}]$ (10), $[3, \sqrt{27}]^*$ (1), $[6, 3+\sqrt{27}]^*$ (2)

$\mathbf{A} = 28.$ 0 (6), 1 (6), 2 (9), 3 (2), 4 (2)
$[1, \sqrt{28}]$ (6), $[28, \sqrt{28}]$ (12), $[2, \sqrt{28}]^*$ (2), $[14, \sqrt{28}]^*$ (5)

$\mathbf{A} = 29.$ 0 (2), 1 (9), 2 (4), 3 (4), 4 (2)
$[1, \sqrt{29}]$ $+$ (16), $[2, 1+\sqrt{29}]^*$ $-$ (5)

$\mathbf{A} = 30.$ 0 (8), 1 (2), 2 (4), 3 (4), 4 (2)
$[1, \sqrt{30}]$ $+++$ (2), $[30, \sqrt{30}]$ $-+-$ (10), $[2, \sqrt{30}]$ $--+$ (2),
$[10, \sqrt{30}]$ $+--$ (6)

$\mathbf{A} = 31.$ 0 (2), 1 (13), 2 (4), 3 (4), 4 (2)
$[1, \sqrt{31}]$ $++$ (8), $[31, \sqrt{31}]$ $--$ (17)

$\mathbf{A} = 32.$ 0 (6), 1 (2), 2 (7), 3 (2), 4 (3)
$[1, \sqrt{32}]$ (2), $[32, \sqrt{32}]$ (11), $[2, \sqrt{32}]^*$ (1), $[4, \sqrt{32}]^*$ (2),
$[16, \sqrt{32}]^*$ (4)

$\mathbf{A} = 33.$ 0 (4), 1 (9), 2 (2), 3 (7), 4 (2)
$[1, \sqrt{33}]$ $++$ (2), $[33, \sqrt{33}]$ $--$ (12), $[2, 1+\sqrt{33}]^*$ $--$ (4),
$[4, 1+\sqrt{33}]^*$ $++$ (6)

$\mathbf{A} = 34.$ 0 (4), 1 (6), 2 (10), 3 (2), 4 (4)
$[1, \sqrt{34}]$ ++ (2), $[34, \sqrt{34}]$ ++ (14), $[3, 1 + \sqrt{34}]$ −− (5),
$[3, 2 + \sqrt{34}]$ −− (5)

$\mathbf{A} = 35.$ 0 (4), 1 (5), 2 (2), 3 (4), 4 (2), 5 (1)
$[1, \sqrt{35}]$ + + + (1), $[35, \sqrt{35}]$ + − − (10), $[5, \sqrt{35}]$ − − + (5),
$[7, \sqrt{35}]$ − + − (2)

$\mathbf{A} = 37.$ 0 (2), 1 (15), 2 (4), 3 (6), 4 (2), 5 (2)
$[1, \sqrt{37}]$ + (12), $[7, 3 + \sqrt{37}]$ + (6), $[7, 4 + \sqrt{37}]$ + (6),
$[2, 1 + \sqrt{37}]^*$ − (7)

$\mathbf{A} = 38.$ 0 (4), 1 (2), 2 (4), 3 (2), 4 (4), 5 (2)
$[1, \sqrt{38}]$ ++ (6), $[38, \sqrt{38}]$ −− (12)

$\mathbf{A} = 39.$ 0 (4), 1 (5), 2 (6), 3 (7), 4 (2), 5 (2)
$[1, \sqrt{39}]$ + + + (4), $[39, \sqrt{39}]$ − + − (12), $[2, 1 + \sqrt{39}]$ − − + (3),
$[6, 3 + \sqrt{39}]$ + − − (7)

$\mathbf{A} = 40.$ 0 (8), 1 (6), 2 (11), 3 (2), 4 (5), 5 (2)
$[1, \sqrt{40}]$ (3), $[40, \sqrt{40}]$ (12), $[5, \sqrt{40}]$ (6), $[8, \sqrt{40}]$ (3),
$[2, \sqrt{40}]^*$ (6), $[10, \sqrt{40}]^*$ (4)

$\mathbf{A} = 41.$ 0 (2), 1 (13), 2 (2), 3 (6), 4 (2), 5 (2)
$[1, \sqrt{41}]$ + (16), $[2, 1 + \sqrt{41}]^*$ + (11)

$\mathbf{A} = 42.$ 0 (8), 1 (2), 2 (4), 3 (4), 4 (4), 5 (2)
$[1, \sqrt{42}]$ + + + (2), $[42, \sqrt{42}]$ − − + (12), $[2, \sqrt{42}]$ − + − (4),
$[3, \sqrt{42}]$ + − − (6)

$\mathbf{A} = 43.$ 0 (2), 1 (13), 2 (4), 3 (4), 4 (4), 5 (2)
$[1, \sqrt{43}]$ ++ (13), $[43, \sqrt{43}]$ −− (16)

$\mathbf{A} = 44.$ 0 (6), 1 (2), 2 (11), 3 (4), 4 (4), 5 (2)
$[1, \sqrt{44}]$ (4), $[44, \sqrt{44}]$ (16), $[2, \sqrt{44}]^*$ (3), $[22, \sqrt{44}]^*$ (6)

$\mathbf{A} = 45.$ 0 (6), 1 (9), 2 (2), 3 (9), 4 (2), 5 (3)
$[1, \sqrt{45}]$ (4), $[45, \sqrt{45}]$ (16), $[3, \sqrt{45}]^*$ (4), $[2, 1 + \sqrt{45}]^*$ (1),
$[6, 3 + \sqrt{45}]^*$ (1), $[10, 5 + \sqrt{45}]^*$ (5)

$\mathbf{A} = 46.$ 0 (4), 1 (10), 2 (10), 3 (2), 4 (6), 5 (2)
$[1, \sqrt{46}]$ ++ (8), $[46, \sqrt{46}]$ −− (26)

$\mathbf{A} = 47.$ 0 (2), 1 (5), 2 (2), 3 (4), 4 (2), 5 (4)
$[1, \sqrt{47}]$ ++ (2), $[47, \sqrt{47}]$ −− (17)

$\mathbf{A} = 48.$ 0 (10), 1 (2), 2 (7), 3 (4), 4 (5), 5 (2), 6 (1)
$[1, \sqrt{48}]$ (1), $[48, \sqrt{48}]$ (12), $[3, \sqrt{48}]$ (3), $[16, \sqrt{48}]$ (4),
$[2, \sqrt{48}]^*$ (2), $[4, \sqrt{48}]^*$ (1), $[6, \sqrt{48}]^*$ (6), $[12, \sqrt{48}]^*$ (2)

$\mathbf{A} = 50.$ 0 (6), 1 (4), 2 (4), 3 (2), 4 (4), 5 (2), 6 (2)
$[1, \sqrt{50}]$ (14), $[2, \sqrt{50}]$ (8), $[5, \sqrt{50}]^*$ (2)

$\mathbf{A} = 51.$ 0 (4), 1 (9), 2 (2), 3 (9), 4 (2), 5 (4), 6 (2)
$[1, \sqrt{51}] + + +$ (7), $[51, \sqrt{51}] - + -$ (14), $[3, \sqrt{51}] + - -$ (5),
$[17, \sqrt{51}] - - +$ (6)

$\mathbf{A} = 52.$ 0 (6), 1 (6), 2 (2), 3 (13), 4 (8), 5 (2), 6 (2)
$[1, \sqrt{52}]$ (10), $[52, \sqrt{52}]$ (16), $[2, \sqrt{52}]^*$ (10), $[4, \sqrt{52}]^*$ (3)

$\mathbf{A} = 53.$ 0 (2), 1 (9), 2 (4), 3 (6), 4 (2), 5 (4), 6 (2)
$[1, \sqrt{53}] +$ (22), $[2, 1 + \sqrt{53}]^* -$ (7)

$\mathbf{A} = 54.$ 0 (8), 1 (2), 2 (8), 3 (6), 4 (4), 5 (2), 6 (2)
$[1, \sqrt{54}]$ (10), $[54, \sqrt{54}]$ (16), $[3, \sqrt{54}]^*$ (2), $[6, \sqrt{54}]^*$ (4)

$\mathbf{A} = 55.$ 0 (4), 1 (13), 2 (4), 3 (4), 4 (4), 5 (5), 6 (2)
$[1, \sqrt{55}] + + +$ (4), $[55, \sqrt{55}] + - -$ (16), $[2, 1 + \sqrt{55}] - - +$ (8),
$[3, 1 + \sqrt{55}] - + -$ (8)

$\mathbf{A} = 56.$ 0 (8), 1 (6), 2 (7), 3 (2), 4 (7), 5 (2), 6 (2)
$[1, \sqrt{56}]$ (2), $[56, \sqrt{56}]$ (14), $[4, 2 + \sqrt{56}]$ (4), $[5, 1 + \sqrt{56}]$ (4),
$[2, \sqrt{56}]^*$ (2), $[14, \sqrt{56}]^*$ (8)

$\mathbf{A} = 57.$ 0 (4), 1 (13), 2 (2), 3 (11), 4 (2), 5 (4), 6 (2)
$[1, \sqrt{57}] ++$ (6), $[57, \sqrt{57}] --$ (16), $[2, 1 + \sqrt{57}]^* --$ (7),
$[6, 3 + \sqrt{57}]^* ++$ (9)

$\mathbf{A} = 58.$ 0 (4), 1 (6), 2 (10), 3 (4), 4 (6), 5 (4), 6 (2)
$[1, \sqrt{58}] ++$ (20), $[2, \sqrt{58}] --$ (16)

$\mathbf{A} = 59.$ 0 (2), 1 (5), 2 (6), 3 (6), 4 (2), 5 (4), 6 (2)
$[1, \sqrt{59}] ++$ (9), $[59, \sqrt{59}] --$ (18)

$\mathbf{A} = 60.$ 0 (12), 1 (2), 2 (11), 3 (4), 4 (6), 5 (2), 6 (3)
$[1, \sqrt{60}]$ (2), $[60, \sqrt{60}]$ (16), $[3, \sqrt{60}]$ (4), $[5, \sqrt{60}]$ (6),
$[2, \sqrt{60}]^*$ (1), $[6, \sqrt{60}]^*$ (3), $[10, \sqrt{60}]^*$ (2), $[30, \sqrt{60}]^*$ (6)

$\mathbf{A} = 61.$ 0 (2), 1 (5), 2 (6), 3 (6), 4 (2), 5 (4), 6 (2)
$[1, \sqrt{61}] +$ (36), $[2, 1 + \sqrt{61}]^* -$ (11)

$\mathbf{A} = 62.$ 0 (4), 1 (2), 2 (4), 3 (2), 4 (4), 5 (2), 6 (4)
$[1, \sqrt{62}]$ ++ (2), $[62, \sqrt{62}]$ −− (20)

$\mathbf{A} = 63.$ 0 (6), 1 (5), 2 (2), 3 (9), 4 (2), 5 (4), 6 (2), 7 (1)
$[1, \sqrt{63}]$ (1), $[63, \sqrt{63}]$ (14), $[7, \sqrt{63}]$ (7), $[2, 1+\sqrt{63}]$ (2), $[3, \sqrt{63}]^*$ (2), $[21, \sqrt{63}]^*$ (5)

$\mathbf{A} = 65.$ 0 (4), 1 (11), 2 (2), 3 (10), 4 (2), 5 (5), 6 (2), 7 (2)
$[1, \sqrt{65}]$ ++ (16), $[5, \sqrt{65}]$ −− (6), $[2, 1+\sqrt{65}]^*$ −− (9), $[10, 5+\sqrt{65}]^*$ ++ (7)

$\mathbf{A} = 66.$ 0 (8), 1 (6), 2 (4), 3 (4), 4 (6), 5 (2), 6 (4), 7 (2)
$[1, \sqrt{66}]$ + + + (8), $[66, \sqrt{66}]$ − − + (16), $[3, \sqrt{66}]$ − + − (4), $[6, \sqrt{66}]$ + − − (8)

$\mathbf{A} = 67.$ 0 (2), 1 (13), 2 (8), 3 (4), 4 (4), 5 (6), 6 (2), 7 (2)
$[1, \sqrt{67}]$ ++ (19), $[67, \sqrt{67}]$ −− (22)

$\mathbf{A} = 68.$ 0 (6), 1 (2), 2 (9), 3 (2), 4 (6), 5 (2), 6 (4), 7 (2)
$[1, \sqrt{68}]$ (4), $[68, \sqrt{68}]$ (16), $[2, \sqrt{68}]^*$ (8), $[4, 2+\sqrt{68}]^*$ (5)

$\mathbf{A} = 69.$ 0 (4), 1 (9), 2 (6), 3 (13), 4 (2), 5 (6), 6 (2), 7 (2)
$[1, \sqrt{69}]$ ++ (8), $[69, \sqrt{69}]$ −− (26), $[2, 1+\sqrt{69}]^*$ −− (2), $[34, 1+\sqrt{69}]^*$ +− (8)

$\mathbf{A} = 70.$ 0 (8), 1 (6), 2 (10), 3 (2), 4 (8), 5 (4), 6 (4), 7 (2)
$[1, \sqrt{70}]$ + + + (6), $[70, \sqrt{70}]$ + − − (18), $[2, \sqrt{70}]$ − + − (10), $[10, \sqrt{70}]$ − − + (10)

$\mathbf{A} = 71.$ 0 (2), 1 (13), 2 (2), 3 (4), 4 (4), 5 (4), 6 (2), 7 (2)
$[1, \sqrt{71}]$ ++ (6), $[71, \sqrt{71}]$ −− (27)

$\mathbf{A} = 72.$ 0 (12), 1 (2), 2 (7), 3 (8), 4 (7), 5 (2), 6 (5), 7 (2)
$[1, \sqrt{72}]$ (2), $[72, \sqrt{72}]$ (16), $[8, 4+\sqrt{72}]$ (5), $[4, 2+\sqrt{72}]$ (3), $[2, \sqrt{72}]^*$ (4), $[3, \sqrt{72}]^*$ (1), $[4, \sqrt{72}]^*$ (8), $[6, \sqrt{72}]^*$ (2), $[24, \sqrt{72}]^*$ (4)

$\mathbf{A} = 73.$ 0 (2), 1 (21), 2 (4), 3 (8), 4 (4), 5 (8), 6 (2), 7 (2)
$[1, \sqrt{73}]$ + (30), $[2, 1+\sqrt{73}]^*$ + (21)

$\mathbf{A} = 74.$ 0 (4), 1 (2), 2 (12), 3 (4), 4 (4), 5 (2), 6 (4), 7 (2)
$[1, \sqrt{74}]$ ++ (20), $[2, \sqrt{74}]$ −− (14)

$\mathbf{A} = 75.$ 0 (6), 1 (5), 2 (2), 3 (9), 4 (2), 5 (5), 6 (4), 7 (2)
$[1, \sqrt{75}]$ (2), $[75, \sqrt{75}]$ (17), $[3, \sqrt{75}]$ (2), $[25, \sqrt{75}]$ (11),
$[5, \sqrt{75}]^*$ (1), $[15, \sqrt{75}]^*$ (2)

$\mathbf{A} = 76.$ 0 (6), 1 (10), 2 (17), 3 (2), 4 (12), 5 (4), 6 (4), 7 (2)
$[1, \sqrt{76}]$ (14), $[76, \sqrt{76}]$ (26), $[2, \sqrt{75}]^*$ (7), $[4, 2+\sqrt{76}]^*$ (10)

$\mathbf{A} = 77.$ 0 (4), 1 (9), 2 (2), 3 (6), 4 (2), 5 (6), 6 (2), 7 (3)
$[1, \sqrt{77}]$ $++$ (4), $[77, \sqrt{77}]$ $--$ (22), $[2, 1+\sqrt{77}]^*$ $++$ (1),
$[14, 7+\sqrt{77}]^*$ $-+$ (7)

$\mathbf{A} = 78.$ 0 (8), 1 (6), 2 (4), 3 (4), 4 (4), 5 (2), 6 (6), 7 (2)
$[1, \sqrt{78}]$ $+++$ (2), $[78, \sqrt{78}]$ $-+-$ (20), $[2, \sqrt{78}]$ $--+$ (4),
$[13, \sqrt{78}]$ $+--$ (10)

$\mathbf{A} = 79.$ 0 (2), 1 (13), 2 (8), 3 (10), 4 (6), 5 (6), 6 (2), 7 (4)
$[1, \sqrt{79}]$ $++$ (2), $[79, \sqrt{79}]$ $--$ (23), $[3, 1+\sqrt{79}]$ $--$ (5),
$[3, 2+\sqrt{79}]$ $--$ (5), $[9, 4+\sqrt{79}]$ $++$ (8), $[9, 5+\sqrt{79}]$ $++$ (8)

$\mathbf{A} = 80.$ 0 (10), 1 (2), 2 (7), 3 (2), 4 (6), 5 (4), 6 (4), 7 (2), 8 (1)
$[1, \sqrt{80}]$ (1), $[80, \sqrt{80}]$ (16), $[5, \sqrt{80}]$ (2), $[16, \sqrt{80}]$ (5),
$[2, \sqrt{80}]^*$ (2), $[4, \sqrt{80}]$ (4), $[8, \sqrt{80}]^*$ (8)

$\mathbf{A} = 82.$ 0 (4), 1 (8), 2 (10), 3 (2), 4 (8), 5 (4), 6 (4), 7 (2), 8 (2)
$[1, \sqrt{82}]$ $++$ (18), $[2, \sqrt{82}]$ $++$ (10), $[3, 1+\sqrt{82}]$ $--$ (8),
$[3, 2+\sqrt{82}]$ $--$ (8)

$\mathbf{A} = 83.$ 0 (2), 1 (5), 2 (2), 3 (4), 4 (2), 5 (4), 6 (2), 7 (4), 8 (2)
$[1, \sqrt{83}]$ $++$ (9), $[83, \sqrt{83}]$ $--$ (18)

$\mathbf{A} = 84.$ 0 (12), 1 (2), 2 (15), 3 (6), 4 (6), 5 (2), 6 (7), 7 (2), 8 (2)
$[1, \sqrt{84}]$ (6), $[84, \sqrt{84}]$ (18), $[7, \sqrt{84}]$ (6), $[12, \sqrt{84}]$ (6),
$[2, \sqrt{84}]^*$ (4), $[6, \sqrt{84}]^*$ (10), $[4, 2+\sqrt{84}]^*$ (1), $[12, 6+\sqrt{84}]^*$ (3)

$\mathbf{A} = 85.$ 0 (4), 1 (21), 2 (6), 3 (6), 4 (4), 5 (11), 6 (2), 7 (4), 8 (2)
$[1, \sqrt{85}]$ $++$ (28), $[5, \sqrt{85}]$ $--$ (18), $[2, 1+\sqrt{85}]^*$ $-+$ (9),
$[10, 5+\sqrt{85}]^*$ $+-$ (5)

$\mathbf{A} = 86.$ 0 (4), 1 (6), 2 (4), 3 (6), 4 (8), 5 (2), 6 (4), 7 (2), 8 (2)
$[1, \sqrt{86}]$ $++$ (16), $[86, \sqrt{86}]$ $--$ (22)

$\mathbf{A} = 87.$ 0 (4), 1 (5), 2 (2), 3 (9), 4 (2), 5 (4), 6 (4), 7 (4), 8 (2)
$[1, \sqrt{87}]$ $+++$ (3), $[87, \sqrt{87}]$ $-+-$ (18), $[3, \sqrt{87}]$ $+--$ (9),
$[29, \sqrt{87}]$ $--+$ (6)

A $= 88.$ 0 (8), 1 (6), 2 (17), 3 (2), 4 (13), 5 (4), 6 (6), 7 (2), 8 (2)
$[1, \sqrt{88}]$ (5), $[88, \sqrt{88}]$ (20), $[8, \sqrt{88}]$ (13), $[11, \sqrt{88}]$ (4),
$[2, \sqrt{88}]^*$ (6), $[4, \sqrt{88}]^*$ (12)

A $= 89.$ 0 (2), 1 (13), 2 (6), 3 (12), 4 (2), 5 (6), 6 (2), 7 (4), 8 (2)
$[1, \sqrt{89}]$ + (28), $[2, 1 + \sqrt{89}]^*$ + (21)

A $= 90.$ 0 (12), 1 (2), 2 (4), 3 (6), 4 (4), 5 (4), 6 (6), 7 (2), 8 (2)
$[1, \sqrt{90}]$ (2), $[90, \sqrt{90}]$ (18), $[2, \sqrt{90}]$ (2), $[18, \sqrt{90}]$ (10),
$[3, \sqrt{90}]^*$ (6), $[6, \sqrt{90}]^*$ (4)

A $= 91.$ 0 (4), 1 (21), 2 (4), 3 (4), 4 (6), 5 (8), 6 (2), 7 (5), 8 (2)
$[1, \sqrt{91}]$ (12), $[91, \sqrt{91}]$ (21), $[7, \sqrt{91}]$ (10), $[13, \sqrt{91}]$ (13)

A $= 92.$ 0 (6), 1 (6), 2 (11), 3 (2), 4 (6), 5 (2), 6 (6), 7 (2), 8 (2)
$[1, \sqrt{92}]$ (6), $[92, \sqrt{92}]$ (24), $[2, \sqrt{92}]^*$ (2), $[46, \sqrt{92}]^*$ (11)

A $= 93.$ 0 (4), 1 (9), 2 (2), 3 (15), 4 (4), 5 (6), 6 (4), 7 (4), 8 (2)
$[1, \sqrt{93}]$ $++$ (10), $[93, \sqrt{93}]$ $--$ (28), $[2, 1 + \sqrt{93}]^*$ $-+$ (3),
$[6, 3 + \sqrt{93}]^*$ $+-$ (9)

A $= 94.$ 0 (4), 1 (6), 2 (18), 3 (4), 4 (8), 5 (4), 6 (4), 7 (4), 8 (2)
$[1, \sqrt{94}]$ $++$ (12), $[94, \sqrt{94}]$ $--$ (42)

A $= 95.$ 0 (4), 1 (5), 2 (6), 3 (4), 4 (2), 5 (7), 6 (2), 7 (4), 8 (2)
$[1, \sqrt{95}]$ $+++$ (2), $[95, \sqrt{95}]$ $+--$ (20), $[2, 1 + \sqrt{95}]$ $--+$ (4),
$[7, 5 + \sqrt{95}]$ $-+-$ (10)

A $= 96.$ 0 (12), 1 (6), 2 (7), 3 (4), 4 (11), 5 (2), 6 (9), 7 (2), 8 (3)
$[1, \sqrt{96}]$ (2), $[96, \sqrt{96}]$ (21), $[32, \sqrt{96}]$ (7), $[2, \sqrt{96}]^*$ (1),
$[3, \sqrt{96}]^*$ (6), $[4, \sqrt{96}]^*$ (2), $[6, \sqrt{96}]^*$ (2), $[8, \sqrt{96}]^*$ (4),
$[16, \sqrt{96}]^*$ (3), $[48, \sqrt{96}]^*$ (8)

A $= 97.$ 0 (2), 1 (21), 2 (4), 3 (8), 4 (6), 5 (10), 6 (2), 7 (6), 8 (2)
$[1, \sqrt{97}]$ + (36), $[2, 1 + \sqrt{97}]^*$ + (25)

A $= 98.$ 0 (6), 1 (2), 2 (4), 3 (2), 4 (4), 5 (2), 6 (4), 7 (2), 8 (4)
$[1, \sqrt{98}]$ (2), $[98, \sqrt{98}]$ (26), $[7, \sqrt{98}]^*$ (2)

A $= 99.$ 0 (6), 1 (9), 2 (6), 3 (15), 4 (2), 5 (4), 6 (4), 7 (4), 8 (2), 9 (1)
$[1, \sqrt{99}]$ (1), $[99, \sqrt{99}]$ (18), $[9, \sqrt{99}]$ (9), $[11, \sqrt{99}]$ (2),
$[7, 1 + \sqrt{99}]$ (4), $[7, 6 + \sqrt{99}]$ (4), $[14, 1 + \sqrt{99}]$ (3),
$[14, 13 + \sqrt{99}]$ (3), $[3, \sqrt{99}]^*$ (3), $[33, \sqrt{99}]^*$ (6)

$\mathbf{A} = 101.$ 0 (2), 1 (15), 2 (2), 3 (6), 4 (4), 5 (6), 6 (4), 7 (4), 8 (2), 9 (2)
$[1, \sqrt{101}]$ + (20), $[4, 1 + \sqrt{101}]$ + (8), $[4, 3 + \sqrt{101}]$ + (8),
$[2, 1 + \sqrt{101}]^*$ − (11)

$\mathbf{A} = 102.$ 0 (8), 1 (2), 2 (8), 3 (4), 4 (4), 5 (4), 6 (6), 7 (2), 8 (4), 9 (2)
$[1, \sqrt{102}]$ + + + (10), $[102, \sqrt{102}]$ − + − (20),
$[3, \sqrt{102}]$ − − + (6), $[6, \sqrt{102}]$ + − − (8)

$\mathbf{A} = 103.$ 0 (2), 1 (13), 2 (8), 3 (4), 4 (4), 5 (8), 6 (2), 7 (6), 8 (2), 9 (2)
$[1, \sqrt{103}]$ ++ (18), $[103, \sqrt{103}]$ − − (33)

$\mathbf{A} = 104.$ 0 (8), 1 (2), 2 (13), 3 (4), 4 (9), 5 (2), 6 (6), 7 (2), 8 (4), 9 (2)
$[1, \sqrt{104}]$ (5), $[104, \sqrt{104}]$ (20), $[8, \sqrt{104}]$ (7), $[13, \sqrt{104}]$ (4),
$[2, \sqrt{104}]^*$ (10), $[4, \sqrt{104}]^*$ (6)

$\mathbf{A} = 105.$ 0 (8), 1 (13), 2 (2), 3 (15), 4 (2), 5 (9), 6 (4), 7 (5), 8 (2), 9 (2)
$[1, \sqrt{105}]$ + + + (4), $[105, \sqrt{105}]$ − + − (20), $[3, \sqrt{105}]$ + − − (4),
$[15, \sqrt{105}]$ − − + (8), $[2, 1 + \sqrt{105}]^*$ − − + (3),
$[4, 1 + \sqrt{105}]^*$ + + + (4), $[6, 3 + \sqrt{105}]^*$ − + − (8),
$[10, 5 + \sqrt{105}]^*$ + − −(11)

$\mathbf{A} = 106.$ 0 (4), 1 (14), 2 (10), 3 (2), 4 (14), 5 (4), 6 (6), 7 (4), 8 (4), 9 (2)
$[1, \sqrt{105}]$ ++ (34), $[2, \sqrt{106}]$ −− (30)

$\mathbf{A} = 107.$ 0 (2), 1 (5), 2 (2), 3 (8), 4 (4), 5 (4), 6 (2), 7 (4), 8 (2), 9 (2)
$[1, \sqrt{107}]$ ++ (13), $[107, \sqrt{107}]$ −− (22)

$\mathbf{A} = 108.$ 0 (12), 1 (2), 2 (11), 3 (8), 4 (6), 5 (2), 6 (9), 7 (2), 8 (4), 9 (2)
$[1, \sqrt{108}]$ (6), $[108, \sqrt{108}]$ (26), $[2, \sqrt{108}]^*$ (5), $[3, \sqrt{108}]^*$ (2),
$[6, \sqrt{108}]^*$ (1), $[9, \sqrt{108}]^*$ (6), $[18, \sqrt{108}]^*$ (2), $[54, \sqrt{108}]^*$ (10)

$\mathbf{A} = 109.$ 0 (2), 1 (21), 2 (12), 3,(10), 4 (4), 5 (12), 6 (2), 7 (8), 8 (2), 9 (2)
$[1, \sqrt{109}]$ + (58), $[2, 1 + \sqrt{109}]^*$ (17)

$\mathbf{A} = 110.$ 0 (8), 1 (2), 2 (4), 3 (2), 4 (4), 5 (4), 6 (4), 7 (2), 8 (4), 9 (2)
$[1, \sqrt{110}]$ + + + (2), $[110, \sqrt{110}]$ + − − (20), $[2, \sqrt{110}]$ − − + (4),
$[5, \sqrt{110}]$ − + − (10)

$\mathbf{A} = 111.$ 0 (4), 1 (13), 2 (2), 3 (9), 4 (4), 5 (4), 6 (6), 7 (4), 8 (2), 9 (2)
$[1, \sqrt{111}]$ + + + (8), $[111, \sqrt{111}]$ − + − (22),
$[2, 1 + \sqrt{111}]$ − − + (5), $[6, 3 + \sqrt{111}]$ + − − (15)

Answers to Exercises

Answers to Exercises for Chapter 1

5. Although the binary algorithm is slightly more efficient, there will probably be little if any difference in the execution times.

6. In binary, 33 becomes 100001 and 21 becomes 10101 so the product in binary is simply 1010110101. This represents 512 + 128 + 32 + 16 + 4 + 1 = 693.

7. 85352841855662246530010669104529797287.

8. The quotient is 1818178099 and the remainder is 3082848597.

Answers to Exercises for Chapter 2

1. If y is a multiple of 3, y^2 is too. If y is one more than a multiple of 3, say $y = 3n + 1$, then $y^2 = 9n^2 + 6n + 1$ is one more than a multiple of 3. If y is two more than a multiple of 3, then $y^2 = (3n + 2)^2 = 9n^2 + 12n + 4$ is again one more than a multiple of 3. Therefore, there is no y whose square is one less than a multiple of 3, let alone one whose square is one less than a *square* times 3.

2. Trial-and-error, best aided by a programmable calculator as in Exercise 9, gives the sequence of ratios $\frac{1}{0}, \frac{1}{1}, \frac{2}{1}, \frac{5}{3}, \frac{7}{4}, \frac{19}{11}, \frac{26}{15}, \frac{71}{41}, \frac{97}{56}, \ldots$. The odd-numbered steps are easily seen to be given by the rules $D = n + d$ and $N = D + 2d$ where n and d are the old numerator and

denominator and N and D are the new ones. The rule $N = D + d$ for the even-numbered steps is also easy to see. The rule for D on even-numbered steps is less obvious. It is (except for the second step) that D is the sum of the two previous denominators, or, in a rule that works in all cases, $D = \frac{n+d}{2}$.

3. The identity $2x^2 + 2(x+y)^2 = (2x+y)^2 + y^2$ is simple algebra. If $2x^2+1 = y^2$, the desired conclusion $2(x+y)^2 - 1 = (2x+y)^2$ follows when $2x^2 + 1$ is subtracted from both sides. Similarly, $2x^2 - 1 = y^2$ implies $(2x + y)^2 - 1 = 2(x + y)^2$.

4. $\frac{1}{0}, \frac{2}{1}, \frac{9}{4}, \frac{38}{17}, \frac{161}{72}, \ldots$ generated by $D = n + 2d$ and $N = 2D + d$ alternates between $n^2 = 5d^2 + 1$ and $n^2 = 5d^2 - 1$.

5. Except for the trivial solution $y = 1$ when $Ax = 0$, the equation $Ax^2 + 1 = y^2$ has a solution if and only if A is not a square, although this is by no means obvious. It is less easy to say when $Ax^2 - 1 = y^2$ has a solution; in the language of Chapter 23, it has a solution if and only if A is not a square and the module $[A, \sqrt{A}]$ is principal (except when $A = x = 1$ and $y = 0$).

7. All solutions of $\square + B = \square$ can be found by taking all factorizations $B = pq$ as a product of two factors in which $p \geq q$ and setting $y = \frac{p+q}{2}$, $x = \frac{p-q}{2}$, ignoring cases in which there are no such numbers y and x (i.e., cases in which one of p and q is even and the other is odd). All solutions of $x^2 + B = y^2$ are obtained in this way. The solutions of the given problem are derived from those, if any, in which x is divisible by the square root of A.

8. The sequence of squares $1, 4, 9, 16, \ldots$ is the sequence $1, 1 + 3, 1 + 3 + 5, 1 + 3 + 5 + 7, \ldots$, as is easily seen. The algorithm checks whether N has the form $1 + 3 + 5 + \cdots + (2n - 1)$ for some n.

10. See the solution of Pell's equation in Chapter 19.

11. $s = 3166815962$ and $d = 4478554083$ satisfy $2s^2 + 1 = d^2$ because in both cases the number is 20057446674355970889.

Answers to Exercises for Chapter 3

1. The squares mod 5 are 0, 1, and 4 and the squares mod 14 are $0, 1, 4, 9, 16 \equiv 2, 2 + 9 \equiv 11, 11 + 11 = 22 \equiv 8$, and $8 + 13 = 21 \equiv 7$

are squares mod 14. At the next step, $7 + 15 = 22 \equiv 8$ is a repeat, and all others will be repeats. A possible algorithm is the following variation on the algorithm of Exercise 8, Chapter 2:

> Input: A modulus a and a number N
> Algorithm:
>   Let $k = 1$, $t = N$
>   While $N \geq a$
>   Reduce N by a
>   End
>   While $a > k$ and $t \neq 0$
>   If $t < k$ add a to t
>   Reduce t by k and increase k by 2
>   End
> Output: If $t = 0$ print "$N \equiv (\frac{k-1}{2})^2$ mod a", else print "N is not a square mod a"

2. The numbers that are 2 mod 3 are 2, 5, 8, 11, 14, 17, 20, ... Mod 5 they are 2, 0, 3, 1, 4, 2, 0, 3, 1, ..., from which it is clear that every fifth term is 1 mod 5, so the solutions are 11, 26, 41, ... In short, the solutions of $x \equiv 2 \bmod 3$ and $x \equiv 1 \bmod 5$ are the solutions of $x \equiv 11 \bmod 15$.

3. $x \equiv 17 \bmod 42$.

4. The multiples of 4 mod 15 are 4, 8, 12, $16 \equiv 1$, 5, 9, 13, $17 \equiv 2$, 6, 10, 14, $18 \equiv 3$, ..., from which it is clear that 7 will be reached for the first time on the $1 + 3 \cdot 4 = 13$th step, because 4 is reached on the first step, and 4 steps increase it by 1. Thereafter, 7 will occur every 15th step. In short, $4x \equiv 7 \bmod 15$ if and only if $x \equiv 13 \bmod 15$.

5. $x \equiv 5 \bmod 7$.

6. Let N be a large enough number that $Na \geq n$. (For example, one could take $N = n$.) Then $u = Na - n$ is a number for which $u + n \equiv 0 \bmod a$ and $x = u + m$ satisfies $x + n \equiv u + m + n \equiv m \bmod a$. Therefore, this congruence has a solution. If $x + n \equiv m \bmod a$ and $x' + n \equiv m \bmod a$, then $x \equiv x + 0 \equiv x + u + n \equiv m + u \equiv x' + n + u \equiv x' \bmod a$, so any two solutions are congruent mod a.

7. $2x \equiv 1 \bmod 4$ obviously has no solution, because an even number cannot be congruent to an odd number mod 4. On the other

hand, $2x \equiv 2 \bmod 4$ has two solutions, 1 and 3, which are not the same mod 4.

8. It is the trivial equivalence relation; any two numbers are congruent mod 1.

9 and 10. See the following chapters.

Answers to Exercises for Chapter 4

2. A possible algorithm is:

Input: A list $a_1, a_2, \ldots, a_n$ of nonzero numbers
Algorithm:
    While $n > 1$
        While $a_1 \neq a_2$
            Set $k = 1$
            While $a_1 > 2ka_2$ or $a_2 > 2ka_2$
                Multiply k by 2
            End
            If $a_1 < a_2$ subtract ka_1 from a_2
            Else subtract ka_2 from a_1
        End
        Drop a_1 and reduce the subscripts on the other entries by 1
    End
Output: The list containing one number a_1 with which the algorithm terminates

(For decimal arithmetic, change 2 to 10 in the three places it occurs.)

3. In the picturesque notation of the question, $\frac{1}{6} \equiv 16 \bmod 19$, $\frac{1}{3} \equiv 13 \bmod 19$, and $\frac{1}{2} \equiv 10 \bmod 19$, so the formula holds. To do the same thing mod 43 requires a solution of the congruence $6x \equiv 1 \bmod 43$. The method of the next chapter finds the solution $x = 36$, which can of course also be found by trial-and-error. Then clearly $u = 2 \cdot 36 \equiv 29 \bmod 43$ satisfies $3u \equiv 1 \bmod 43$ (because $3u = 3 \cdot 2x = 6x$) and $v = 3 \cdot 36 \equiv 22 \bmod 43$ satisfies $2v \equiv 1 \bmod 43$ (because $2v = 6x$). Thus, mod 43, the equation $\frac{1}{6} + \frac{1}{3} = \frac{1}{2}$ becomes the congruence $36 + 29 \equiv 22 \bmod 43$, which is correct. In the mod 143 case, one can similarly find that $x = 24$ is a solution of $6x \equiv 1 \bmod 143$; then $u \equiv 2x \bmod 143$ and $v \equiv 3x \bmod 143$ satisfy $x + u \equiv v \bmod 143$ is described by $\frac{1}{6} + \frac{1}{3} = \frac{1}{2}$ because $3u \equiv 1$ and $2v \equiv 1$. The reason this always works is that $6x \equiv 1 \bmod n$ has exactly one solution

when n is relatively prime to 6 (see the next chapter); then $3u \equiv 1 \bmod n$ and $2v \equiv 1 \bmod n$ have just the solutions $2x$ and $3x \bmod n$ and the statement that $x + u \equiv v \bmod n$ follows immediately from the observation that both sides are $\equiv 3x \bmod n$. This is simply the process of finding the least common denominator 6 of the fractions and writing the equation as the obvious equation $\frac{1}{6} + \frac{2}{6} = \frac{3}{6}$.

Answers to Exercises for Chapter 5

1. (a) $283 \cdot 123 = 76 \cdot 458 + 1$ and $47 \cdot 458 = 175 \cdot 123 + 1$. The algorithm takes 13 steps, or 11 if the "speeded up" version is used.

(b) The number $47 \cdot 458$ is 0 mod 458 and 1 mod 123, while the number $283 \cdot 123$ is 0 mod 123 and 1 mod 458. Therefore, $100 \cdot 47 \cdot 458$ is 0 mod 458 and 100 mod 123 while $300 \cdot 283 \cdot 123$ is 0 mod 123 and 300 mod 458, so the sum of these two numbers solves the given simultaneous congruences.

(c) Division of the solution $100 \cdot 47 \cdot 458 + 300 \cdot 283 \cdot 123$ by $123 \cdot 458$ leaves a remainder 32818, which is the smallest solution.

2. (a) If $x \geq b$ in a solution of $d + xa = yb$, then $yb = d + xa > xa \geq ba$ so $y > a$ and ab can be subtracted from both sides to find a solution $d + (x - b)a = (y - a)b$ in which x is reduced to $x - b$. Repetition of this procedure must eventually reach a solution in which $x < b$. (b) Since the equation $d + xa = yb$ can be divided by d, it will suffice to show that there is at most one solution (x, y) of an equation of the form $1 + xa = yb$ in which $x < b$. If $1 + x'a = y'b$, then $x' \equiv x' + x(1 + x'a) \equiv x' + x + xx'a = x + x'(1 + xa) \equiv x \bmod b$, so if x and x' are both less than b, they must be equal.

3. (a) At each step of the new algorithm, define d to be $va - ub$ and e to be $yb - xa$. Then both algorithms terminate if and only if $d = e$, which is to say, if and only if $a(v + x) = b(u + y)$. Otherwise, both algorithms leave x and y unchanged and change u to $u + y$ and v to $v + x$ when $d > e$—which is to say when $a(v + x) > b(u + y)$—and both leave u and v unchanged and change y to $u + y$ and x to $v + x$ when $d < e$—which is to say when $a(v + x) < b(u + y)$.

(b) Multiply the inequalities $\frac{u}{v} < \frac{p}{q} < \frac{y}{x}$ by qxv to restate them as $uqx < pxv < yqv$. Let this be written $A < B < C$ where $A = uqx$, $B = pxv$, and $C = yqv = q(ux + 1) = A + q$. Since A and B are both multiples of x, $B - A \geq x$. Therefore, $q = C - A > B - A \geq x$. Similarly, B and C are multiples of v, so $q = C - A > C - B \geq v$. Thus, $q > x$ and $q > v$ as was to be shown.

(c) When the algorithm terminates, $\frac{a}{b} = \frac{u+y}{v+x}$. Moreover $xu+1 = yv$ at each step of the algorithm because this is true of the initial values $x = u = 0$ and $y = v = 1$ and it remains true when u changes to $u+y$ and v changes to $v+x$ (because $x(u+y)+1 = xu+xy+1 = yv + xy = y(v + x)$) and when y changes to $u + y$ and x changes to $v + x$ (because $(v + x)u + 1 = vu + xu + 1 = vu + vy = (u + y)v)$. Moreover, $ux < yv$ implies $uvx+ux^2 < uvx+yvx < yv^2+yvx$, which implies $\frac{u}{v} < \frac{a}{b} < \frac{y}{x}$ (divide by $vx(v + x)$). Therefore, $b > x$ and $b > v$ by (b). Then, since $\frac{u}{v} < \frac{a}{b}$, which is to say $ub < va$, it follows that $uv < ub < av$, so $u < a$.

4. According to Exercise 1, one can assume that $u < a$. Then $d + ub = va$ can be subtracted from $ab + d = ab + d$ to find $b(a - u) = d + a(b - v)$, which shows that $b \geq v$ and $x = b - v$ and $y = a - u$. In short, one of (u, v) and (x, y) determines the other via $v + x = b$ and $u + y = a$.

5. When a and b are relatively prime, d is 1 and (1) implies $yb \equiv 1 \bmod a$. In the suggestive notation of Exercise 3, Chapter 4, $y \equiv \frac{1}{b} \bmod a$.

6 and 7. Multiplication of $ax \equiv c \bmod b$ by a reciprocal v of a mod b gives $vax \equiv vc \bmod b$, so $x \equiv vc \bmod b$ for any solution x of $ax \equiv c \bmod b$. Conversely, $x = vc$ satisfies $ax \equiv avc \equiv c \bmod b$.

8. A possible formulation of the algorithm is:

```
Input: Three numbers a, b, and c, with b ≥ a > 0.
Algorithm:
    t = 0
    While t = 0
      m = 1
      While (m + 1)a ≤ b
        m = m + 1
      End
      If ma = b then t = 1
```

Else $a = (m+1)a - b$ and $c = (m+1)c$
End
Let q be the quotient when c is divided by a, and let r be the remainder
Output: If $r \neq 0$ print "no solution" else print "$x \equiv q \bmod m$"

(One might choose to reduce c mod b each time it is changed, or at least to reduce q mod m at the last step. Also, instead of adding 1 to m, one might add the largest power of 2 that can be added without violating $ma \leq b$.)

10. $m = 3$ because the 4-digit number $1369 = 37^2$ is the smallest composite number whose factors are all greater than 31.

Answers to Exercises for Chapter 6

1. The simultaneous congruences $x \equiv 2 \bmod 3$, $x \equiv 3 \bmod 5$, and $x \equiv 2 \bmod 7$ are equivalent to the single congruence $x \equiv 23 \bmod 3{\cdot}5{\cdot}7$, the most general solution of which is $23 + 105i$, where $i = 0, 1, 2, \ldots$.

2. The numbers $k = 0, 1, 2, \ldots, 34$, in that order, correspond to the pairs $(0,0)$, $(1,1)$ $(2,2)$, $(3,3)$, $(4,4)$, $(0,5)$, $(1,6)$, $(2,0)$, $(3,1)$, $\ldots$, $(3,5)$, $(4,6)$, in that order, where the step from one pair to the next adds 1 to each entry, except that it reduces the first entry to 0 when it is 4 and reduces the second entry to 0 when it is 6.

3. There are 35 letters and 35 pigeonholes. If at least one letter goes into each pigeonhole, then no two letters can go into the same pigeonhole.

4. If $a_i = 0$, then the condition that a_i must be relatively prime to each a_j for $j \neq i$ implies that $a_j = 1$ for $j \neq i$. Therefore, the congruences $x \equiv m_j \bmod a_j$ impose no condition on x at all for $j \neq i$, whereas $x \equiv m_i \bmod a_i$ is the condition $x = m_i$. This is the same as a single condition $x \equiv m_i \bmod a_1 a_2 \cdots a_l$, as the Chinese remainder theorem says it should be.

5. (a) Since $x \equiv m \bmod a$ and $x \equiv n \bmod b$ imply $x \equiv m \bmod d$ and $x \equiv n \bmod d$, the existence of a solution x implies $m \equiv n \bmod d$. Conversely, if $m \equiv n \bmod d$, then $x \equiv 0 \bmod \frac{a}{d}$ and $x \equiv \frac{n-m}{d} \bmod \frac{b}{d}$ has a solution (assume without loss of generality that $m \leq n$) because $\frac{a}{d}$ and $\frac{b}{d}$ are relatively prime. (Division of $d + ua = vb$ by d gives an equation which shows that $\frac{a}{d}$ and $\frac{b}{d}$ are relatively prime.) For such

an x, $X = xd + m$ satisfies $X \equiv m \bmod a$ and $X \equiv n \bmod b$. (b) Clearly if x satisfies the given congruences, then $x' = x + k \cdot \frac{ab}{d}$ also satisfies them for every k. Conversely, if x and x' are solutions of $x \equiv m \bmod a$ and $x \equiv n \bmod b$, their difference, call it y, is a solution of $y \equiv 0 \bmod a$ and $y \equiv 0 \bmod b$. Thus $y \equiv 0 \bmod \frac{a}{d}$ and $\frac{y}{d} \equiv 0 \bmod \frac{b}{d}$. Since 0 is a solution of these congruences, the theorem of the text then implies that $\frac{y}{d} \equiv 0 \bmod (\frac{a}{d}) \cdot (\frac{b}{d})$, so y is a multiple of $d \cdot (\frac{a}{d}) \cdot (\frac{b}{d}) = \frac{ab}{d}$, as was to be shown.

6. The solution of $q \cdot 14 + 10 \equiv 20 \bmod 23$ can be accomplished by the simple steps $14q \equiv 10 \bmod 23$, $28q \equiv 20 \bmod 23$, $5q \equiv 20 \bmod 23$, from which the solution $q = 4$ is apparent and the solution $4 \cdot 14 + 10 = 66$ follows.

Answers to Exercises for Chapter 7

2. What is to be shown is that if $Ax^2 = y^2$ where A, x, and y are numbers, then A is a square. The fundamental theorem of arithmetic implies that the number of times any given prime factor divides a square is even. Therefore, for any prime factor p, the number of times p divides A plus an even number is equal to an even number. Therefore, the number of times p divides A is even. Therefore A is the square of the number that is the product of the same distinct prime factors as A in which each occurs half as many times as it occurs in A.

4 and 5. The improvement of Exercise 4 should, roughly, make the program twice as fast, and that of Exercise 5 should be an even greater improvement.

Answers to Exercises for Chapter 8

1. $a^0 = 1$ when $a \neq 0$ and $0^b = 0$ when $b \neq 0$. There is no reasonable definition of 0^0, any more than there is of $\frac{0}{0}$.

2. The orders of the numbers 1 through 12 mod 13 are 1, 12, 3, 6, 4, 12, 12, 4, 3, 6, 12, 2, respectively.

4. $2^{10} \equiv 1 \bmod 11$, $2^{20} \equiv 4 \bmod 21$, $2^{30} \equiv 1 \bmod 31$, $2^{40} \equiv 1 \bmod 41$, $2^{50} \equiv 4 \bmod 51$, $2^{60} \equiv 1 \bmod 61$, $2^{70} \equiv 1 \bmod 71$, $2^{80} \equiv 40 \bmod$

81, $2^{90} \equiv 64 \bmod 91$, and $2^{100} \equiv 1 \bmod 101$. As for regularities, see later chapters, especially Chapter 11.

6. The theorems of Chapter 10 will provide ways of predicting the outcomes of many such calculations—such as computing $a^{n-1} \bmod n$—which can be used to check that the program you develop is producing correct answers.

Answers to Exercises for Chapter 9

1. A number in the suggested form $qc + r$ is relatively prime to c if and only if r is relatively prime to c. Therefore, exactly $p \cdot \phi(c)$ numbers less than pc are relatively prime to c (q can have any of the values $0, 1, \ldots, p-1$ but r must be relatively prime to c) and the problem is to determine how many of these are relatively prime to pc. If p divides c, then all of them are relatively prime to pc. Otherwise, for any given r, exactly one of the p numbers $qc + r$ is divisible by p because the congruence $qc + r \equiv 0 \bmod p$ has a unique solution $q \bmod p$ (c has a reciprocal mod p). Therefore $\phi(pc) = (p-1)\phi(c)$ in this case.

2. The formula doesn't really involve fractions, because to multiply c by $(1 - \frac{1}{p})$ means to divide c by p (and by assumption p is a divisor of c) and then multiply by $p - 1$. The proposition of Exercise 1 shows that if the formula gives the correct value of $\phi(c)$, then it gives the correct value of $\phi(pc)$ for any prime c. Moreover, it gives the correct value of $\phi(1)$, in which case there are *no* prime divisors of c. Therefore, the value of $\phi(c)$ for any product of primes c can be found starting with $\phi(p) = p - 1$ for one of the prime factors of c and then multiplying by the primes one at a time. Each time c is multiplied by a *new* prime, $\phi(c)$ is multiplied by $p - 1$ instead of p.

3. $\phi(60) = 16$. It counts 1, the 14 primes greater than 5 and less than 60, and the composite number 49.

4. The formula of Exercise 2 shows that both $\phi(mn)$ and $\phi(m)\phi(n)$ are $mn(1 - \frac{1}{p_1})(1 - \frac{1}{p_2}) \cdots (1 - \frac{1}{p_k})$ where $p_1, p_2, \ldots, p_k$ are the distinct prime factors of mn, provided m and n have no prime factors in common.

Answers to Exercises for Chapter 10

2. (a) For any numbers a and b and for any prime p, $(a+b)^p \equiv a^p + b^p \bmod p$.

(b) The coefficient of $a^k b^{p-k}$ in the expansion of $(a+b)^p$ is $\frac{p!}{k!(p-k)!}$. The numerator contains the factor p and the denominator does not, unless $k = p$ or $k = 0$ (because p, being prime, does not divide any number less than p). Therefore, this coefficient, which is of course a number, is 0 mod p.

(c) The formula in (a) implies $(a+b+c)^p \equiv (a+b)^p + c^p \equiv a^p + b^p + c^p \bmod p$ and in the same way implies $(a+b+\cdots+e)^p \equiv a^p + b^p + \cdots + e^p \bmod p$ for any number of summands a, b, ..., e. When there are k summands, all equal to 1, this formula says that $k^p \equiv k \bmod p$, which is Fermat's theorem.

3. By the corollary, the order of 2 mod p divides n. Since n is assumed to be prime, the order of 2 mod p must then *be* n. Also by the corollary, the order of 2 mod p divides $\phi(p) = p - 1$, so $p \equiv 1 \bmod n$, as was to be shown.

A prime factor of $2^7 - 1$ would have to be congruent to 1 mod 7. The smallest such prime is 29, which is greater by far than the square root of $2^7 - 1$, so $2^7 - 1$ is prime because it has no prime factors less than its square root.

A prime factor of $2^{11} - 1$ must be 1 mod 11, so the first prime to be tested is 23. In fact, $2^{11} \equiv 1 \bmod 23$ by simple exponentiation, so $2^{11} - 1 = 2047$ has the factor 23 and is not prime.

The only possible prime factors of $2^{13} - 1$ are 53 and 79, because all other primes that are congruent to 1 mod 13 are greater than the square root of $2^{13}-1$. Because $2^{13} \equiv 30 \bmod 53$ and $2^{13} \equiv 55 \bmod 79$, these primes do not divide $2^{13} - 1$, so it must be prime.

4. (a) $(14ed6li9ah)(2873bkfgjc)$. (b) $(1248e7d36blkif9gajhc)$. (d) $(1e6ia)(4dl9h)(27bfj)(83kgc)$. (e) $(16aei)(4lhd9)(2bj7f)(8kc3g)$.

5. The orders of 99, 100, and 101 mod 221 are 16, 24, and 6, respectively.

Answers to Exercises for Chapter 11

1. The answer is not clear. On the one hand, if n passes Miller's test for all prime bases less than 10, for example, then it passes Miller's test for all bases less than 10, and even for all bases that are products of just these primes. On the other hand, if n fails Miller's test for just one of two prime bases, it will fail for their product, and even if it fails for both, it seems less likely to pass for their product; therefore, it would seem that one test with the base 6 would have almost as good a chance of spotting a composite number as two tests, one for the base 2 and one for the base 3. Since the tests require relatively little computation, it is not very important to decide.

2.

```
Input: n, a      (1 < a < n and n is odd)
Algorithm:
    i = 1
    e = (n − 1)/2
    While 2 divides e
      e = e/2
      i = i + 1
    End
    t = 1
    Compute u ≡ a^e mod n
    If u ≡ 1 mod n then t = 0
    While t = 1 and i > 1
      If u ≡ −1 mod n then t = 0
      Else u ≡ u^2 mod n and i = i − 1
    End
Output: If t = 1, n is composite. Otherwise, n passes Miller's test for the
    base a
```

(If $a^{(n-1)/2^i} \equiv 1 \bmod n$, then n passes Miller's test for the base a. Otherwise, n passes Miller's test only if $a^{(n-1)/2^{i-j}} = -1$ for some $j = 0, 1, \ldots, i-1$.)

3. $2^{200} \equiv 1$ but $2^{100} \equiv 2260 \bmod 12801$ and $5^{400} \equiv 1$ but $5^{200} \equiv 2260 \bmod 12801$. (On the other hand $3^{6400} \equiv 2874 \bmod 12801$ and $7^{6400} \equiv 9436 \bmod 12801$, so 12801 fails Miller's test for the bases 3 and 7 immediately.)

Answer to Exercise for Chapter 12

If you tried to do something clever in choosing e or in choosing p and q, you were probably playing into the hands of codebreakers. It is essential to make random choices.

Answers to Exercises for Chapter 13

1. The first 12 powers of 2 mod 13 are distinct. They are 2, 4, 8, $16 \equiv 3$, 6, 12, $24 \equiv 11$, $22 \equiv 9$, $18 \equiv 5$, 10, $20 \equiv 7$, $14 \equiv 1$. There are $\phi(12) = 4$ primitive roots mod 13, namely, 2, $2^5 \equiv 6$, $2^7 \equiv 11$, and $2^{11} \equiv 7$. The remaining numbers less than 13 are partitioned into $\phi(6) = 2$ numbers whose order mod 13 is 6 (they are $2^2 = 4$ and $2^{10} \equiv 10$), $\phi(4) = 2$ numbers whose order is 4 (they are $2^3 = 8$ and $2^9 \equiv 5$), $\phi(3) = 2$ numbers whose order is 3 (they are $2^4 \equiv 3$ and $2^8 \equiv 9$), $\phi(2) = 1$ number whose order is 2 (it is $2^6 \equiv 12 \equiv -1$) and $\phi(1) = 1$ number whose order is 1 (it is 1).

2. Partition the $p - 1$ positive numbers less than p into subsets S_r by putting in S_r all numbers whose orders mod p are r. The subset S_r is nonempty if and only if r divides $p - 1$, and in this case it contains exactly $\phi(r)$ elements. The formula says simply that the $p-1$ numbers in all the sets put together can be counted by counting the elements in each of the sets separately and adding the results.

3. If a has order 25 mod p, then a^5, a^{10}, a^{15}, and a^{20} have order 5 mod p, $a^{25} \equiv 1$ has order 1 mod p, and the remaining 20 distinct powers of a mod p all have order 25.

4. Neither 2^{50} nor 2^{20} is congruent to 1 mod 101, so 2 is a primitive root mod 101. Therefore, its 4th power 16 has order $25 = \frac{101-1}{4}$ mod 101, and $16^5 \equiv 95$ mod 101 has order 5. Exercise 3 then shows how to find all numbers whose orders mod 101 are 5 or 25, which are the other possible answers.

5. Gauss's method in this case is simply to compute 2^{128} mod 257, 3^{128} mod 257, until an answer other than 1 (which must be -1) is found. In fact, $3^{128} \equiv -1$ mod 257, so 3 is a primitive root mod 257. There are 128 primitive roots mod 257, namely, the odd powers of 3. Since $11^{128} \equiv 1$ mod 257, it is not a primitive root mod 257.

7. There are none for 43. There are two each for the other two; for 37 they are ± 6 and for 41 they are ± 9.

Answers to Exercises for Chapter 14

1. (a) The numbers are 12, 0, 11, 0, 0, 11, 11, 11, 11, 0, 0, 11, 0. Simply put, the answer is 0 for nonzero squares mod 13, 12 for 0, and 11 for other numbers. Note that $x^6 - 1$ has 6 roots mod 13.

(b) $x^6 - 1 \equiv (x-1)(x^5 + x^4 + x^3 + x^2 + x + 1) + 0 \bmod 13$. (In fact, this congruence is true mod n for any n.)

(c) The roots, by direct computation using a calculator, are 3 (the value for 3 is $364 \equiv 0 \bmod 13$), 4 ($1365 \equiv 0 \bmod 13$), 9 ($66430 \equiv 0 \bmod 13$), 10 ($111111 \equiv 0 \bmod 13$), and 12 ($271453 \equiv 0 \bmod 13$). Note that this polynomial of degree 5 has 5 roots.

(d) The six roots of $x^6 - 1$ are partitioned by the factorization $(x-1)(x^5 + x^4 + x^3 + x^2 + x + 1)$ into the roots of the first factor (the single root 1) and the roots of the second factor (the other five).

2. By Fermat's theorem, all 13 numbers less than 13 are roots. Therefore, by the theorem of this chapter, $x - a$ divides $x^{13} - x$ for all a and $x^{13} - 1 \equiv x(x-1)(x-2)(x-3)(x-4)(x-5)(x-6)(x-7)(x-8)(x-9)(x-10)(x-11)(x-12) \bmod 13$, a result that can be verified by direct multiplication (somewhat lengthy).

3. The squares mod 7 are 0, 1, 2, and 4. Therefore, $x^2 \equiv 3 \bmod 7$ is impossible, which is to say that $x^2 + 4$ has no root mod 7. Therefore, $x^3 + 4x$ has the single root $x \equiv 0 \bmod 7$. The cubes mod 7 are 0 and ± 1. Therefore, $x^3 + 2$ has no roots mod 7. There are of course many other examples.

Answers to Exercises for Chapter 15

1.

numbers

	0	*1*	*2*	*3*	*4*	*5*	*6*	*7*	*8*	*9*
0		17	10	15	07	26	08	12	18	27
1	25	22	02	03	20	30	14	21	16	24
2	05	23	19	13	04	06	09	29	28	11
3	01									

indices

	0	*1*	*2*	*3*	*4*	*5*	*6*	*7*	*8*	*9*
0		30	12	13	24	20	25	04	06	26
1	02	29	07	23	16	03	18	01	08	22
2	14	17	11	21	19	10	05	09	28	27
3	15									

3. The quadratic formula calls for a square root of $2^2 - 4 \cdot 1 \cdot (-1) = 8$. The index of 8 for $p = 151$ is 60, so 8 has two square roots mod 151, the numbers whose indices are 30 and $30 + \frac{151-1}{2} = 105$, which are the numbers 59 and 92. The roots of x^2+2x-1 mod 151 are $\frac{-2+59}{2}$ and $\frac{-2+92}{2} = 45$. The first number is 104, as follows easily from $\frac{1}{2} \equiv \frac{152}{2} \equiv 76$ or even more easily from $r_1 + r_2 \equiv -2 \equiv 149 \bmod 151$. Indeed, $45^2 + 2 \cdot 45 - 1 = 2114 = 14 \cdot 151$ and $104^2 + 2 \cdot 104 - 1 = 11023 = 73 \cdot 151$. The index of 8 for $p = 157$ is 129; since it is odd, $x^2 + 2x - 1$ has no roots mod 157.

4. $\phi(156) = \phi(2^2 \cdot 3 \cdot 13) = 2 \cdot 2 \cdot 12 = 48$ is the number of primitive roots mod 157. The one used in constructing the table is 139. Two others are $139^5 \equiv 84$ and $139^7 \equiv 55 \bmod 157$ because 5 and 7 are relatively prime to 156. The number of primitive roots mod 151 is $\phi(150) = 40$.

5. The index 122 of 3 is not divisible by 3, so 3 is not a cube mod 157. The index 147 of 2 is $3 \cdot 49$, so 2 has three cube roots, the numbers whose indices are 49, $49 + \frac{156}{3} = 101$, and $49 + \frac{2 \cdot 156}{3} = 153$. These are 62, 136, and 116, respectively. Indeed, $62^3 = 2 + 1518 \cdot 157$, $136^3 \equiv 2 + 16022 \cdot 157$, $116^3 = 2 + 9942 \cdot 157$.

6. The smallest solution j of $30^j \equiv 1 \bmod 157$ is the smallest solution j of $124j \equiv 0 \bmod 156$. Divide by the greatest common divisor 4 of 124 and 156 to find $31j \equiv 0 \bmod 39$, the smallest solution of which is $j = 39$. Thus, 39 is the order of 30 mod 157. The order of 90 mod 157 is the smallest solution j of $90j \equiv 0 \bmod 156$, which is $j = 26$.

7. $10^6 \equiv 67 \bmod 157$ and $10^6 \equiv 78 \bmod 151$. (In both cases the index of 10 is 2 so 10^6 is the number whose index is 12.)

Answers to Exercises for Chapter 16

1. The nth term of the sequence is the ratio in which the denominator is the coefficient of $\sqrt{2}$ in $(1+\sqrt{2})^n$ and the numerator is the other coefficient. (This formula is clearly correct for $n = 1$. If it is correct for $n-1$, then $(1+\sqrt{2})^n = (N_{n-1} + D_{n-1}\sqrt{2})(1+\sqrt{2}) = (N_{n-1} + 2D_{n-1}) + (N_{n-1} + D_{n-1})\sqrt{2}$ where N_{n-1} and D_{n-1} are the numerator and denominator, respectively, of the $(n-1)$st term of the sequence. The definition of the sequence, which is $D_n = N_{n-1} + D_{n-1}$ and $N_n = D_n + D_{n-1} = N_{n-1} + 2D_{n-1}$, then shows that the formula is correct for n.)

2. A term with odd index is found by multiplying the preceding term by $1+\sqrt{3}$. Thus, $N + D\sqrt{3} = (1+\sqrt{3})(n + d\sqrt{3})$, which gives the desired formulas $D = n + d$ and $N = n + 3d = D + 2d$. A term with even index is found by multiplying the preceding term by $\frac{2+\sqrt{3}}{1+\sqrt{3}} = \frac{1+\sqrt{3}}{2}$. Thus $N + D\sqrt{3} = \frac{1}{2}(n + 3d + (n+d)\sqrt{3}) = d + \frac{n+d}{2} + \frac{n+d}{2}\sqrt{3}$, which gives the desired formulas. (For terms with odd index, the difference between numerator and denominator is an even number—it is twice the previous denominator—so their sum is divisible by 2.)

3. The nth term of the sequence (beginning with $n = 0$) has the coefficient of $\sqrt{5}$ in $(2+\sqrt{5})^n$ in its denominator and the other coefficient of $(2+\sqrt{5})^n$ in its numerator, from which the formulas $D = n + 2d$ and $N = 2D + d$ follow easily.

Answers to Exercises for Chapter 17

1. Of course there are many ways to prove these equations. For example:

(a) The formula $y^2 + x\sqrt{A}(y + x\sqrt{A}) = y^2 + xy\sqrt{A} + Ax^2 = y(y + x\sqrt{A}) + Ax^2$ shows that $y^2 \equiv Ax^2 \bmod [y + x\sqrt{A}]$. Therefore, $49 \equiv 44 \bmod [7 + 2\sqrt{11}]$ so $5 \equiv 0 \bmod [7 + 2\sqrt{11}]$ and $[7 + 2\sqrt{11}] = [5, 7 + 2\sqrt{11}]$. This in turn is $[5, 7 + 2\sqrt{11}, 21 + 6\sqrt{11}]$ (the last entry is three times the second). Because $21 + 6\sqrt{11} \equiv 1 + \sqrt{11} \bmod 5$ and $7 + 2\sqrt{11} \equiv 0 \bmod [5, 1 + \sqrt{11}]$, the desired equation follows.

(b) The congruence $y^2 \equiv Ax^2 \bmod [y+x\sqrt{A}]$ in the answer to (a) implies $49 \equiv 7 \bmod [7+\sqrt{7}]$ and $169 \equiv 112 \bmod [13+4\sqrt{7}]$, so $49-7 \equiv 0 \equiv 169-112 \bmod [7+\sqrt{7}, 13+4\sqrt{7}]$ and the given module is equal to $[7+\sqrt{7}, 13+4\sqrt{7}, 42, 57]$. By the Euclidean algorithm, $[42, 57] = [3]$, so the given module is $[3, 7+\sqrt{7}, 13+4\sqrt{7}] = [3, 1+\sqrt{7}, 1+\sqrt{7}]$, as was to be shown.

2. $5\sqrt{11} + 2\sqrt{11}(22 + 7\sqrt{11}) = 7(22 + 7\sqrt{11})$ and $11 + \sqrt{11} + 3(22 + 7\sqrt{11}) = \sqrt{11}(22 + 7\sqrt{11})$.

3. First, the fact that $y^2 \equiv Ax^2 \bmod [y + x\sqrt{A}]$ as in Exercise 1 implies that if $y + x\sqrt{A}$ is a nonzero entry of the list, then $|y^2 - Ax^2|$ is a *nonzero* number that can be annexed to the list. All subsequent calculations can be regarded as calculations mod n for some number n that is in the list. If there is more than one number (as opposed to hypernumbers) in the list, they can be replaced by their greatest common divisor using the Euclidean algorithm. Therefore, one can assume without loss of generality that the list contains just one number, which is not zero. If n is that number, then $n\sqrt{A}$ can be annexed to the list. Therefore, one can assume without loss of generality that the list contains at least one hypernumber that is not a number. If $v + u\sqrt{A}$ and $y + x\sqrt{A}$ are hypernumbers in the list in which $0 < x < u$ and x does not divide u, then a hypernumber in which the coefficient of $\sqrt{A}$ is less than x can be annexed to the list, namely, $mn + (v + u\sqrt{A}) - q(y + x\sqrt{A})$, where q is the quotient when u is divided by x and m is large enough to make the subtraction possible. Therefore, one can assume without loss of generality that the list of hypernumbers has the property that the smallest coefficient of $\sqrt{A}$ among the hypernumbers in the list divides the coefficients of $\sqrt{A}$ in all hypernumbers in the list. Then n and $y + x\sqrt{A}$ can be used to replace all hypernumbers in the list other than n and $y + x\sqrt{A}$ with numbers. Then the Euclidean algorithm can be used to put the module in the form $[n, y + x\sqrt{A}]$ where $x > 0$. Moreover, since $n\sqrt{A}$ and $xA + y\sqrt{A}$ can be annexed to this list, one can also assume x divides both n and y, because otherwise $[n, y + x\sqrt{A}]$ can be replaced with another module in the same form in which x is reduced. In other

words, the given module can be put in the form $[e][f, g+\sqrt{A}]$, as was to be shown. Since this can be written $[e][f, |g^2-A|, g+\sqrt{A}]$, one can also assume $g^2 \equiv A \bmod f$, since otherwise it is equal to one of the same form in which f is reduced. See also the proof in Chapter 18.

4. The given product can be written $[f, \mathcal{G}+\sqrt{A}][F, \mathcal{G}+\sqrt{A}]$ because $[f, \mathcal{G}+\sqrt{A}] = [f, g+\sqrt{A}]$ and $[F, \mathcal{G}+\sqrt{A}] = [F, G+\sqrt{A}]$. By definition, this product module is $[fF, f(\mathcal{G}+\sqrt{A}), F(\mathcal{G}+\sqrt{A}), (\mathcal{G}+\sqrt{A})^2]$. The augmented Euclidean algorithm gives a solution (x, y) of $xf+1 = yF$, because f and F are relatively prime. The hypernumber $yF(\mathcal{G}+\sqrt{A})$ can be annexed to this list of four hypernumbers, and then $f(\mathcal{G}+\sqrt{A})$ can be subtracted from it x times to find that the product module is $[fF, f(\mathcal{G}+\sqrt{A}), F(\mathcal{G}+\sqrt{A}), (\mathcal{G}+\sqrt{A})^2, \mathcal{G}+\sqrt{A}]$. The middle three terms are multiples of the last and can therefore be dropped.

Answers to Exercises for Chapter 18

1. (a) See Exercise 1(a) of Chapter 17. (b) $[7, 2+\sqrt{3}] = [7, 2+\sqrt{3}, 3+2\sqrt{3}] = [7, 2+\sqrt{3}, 10+2\sqrt{3}] = [7, 2+\sqrt{3}, 6] = [1, 2+\sqrt{3}] = [1, \sqrt{3}]$. (c) $[11, 10+2\sqrt{3}] = [11, 10+2\sqrt{3}, 60+12\sqrt{3}] = [11, 10+2\sqrt{3}, 5+\sqrt{3}]$ (reduce coefficients of last entry mod 11) $= [11, 5+\sqrt{3}]$. (d) $[25+6\sqrt{3}, 20+7\sqrt{3}] = [517, 25+6\sqrt{3}, 20+7\sqrt{3}] = [517, 25+6\sqrt{3}, 537+7\sqrt{3}] = [517, 25+6\sqrt{3}, 512+\sqrt{3}] = [517, 5170+25-6\cdot 512, 512+7\sqrt{3}] = [517, 2123, 512+\sqrt{3}] = [11, 512+\sqrt{3}] = [11, 6+\sqrt{3}]$.

2. First, $[e] = [x, y]$. Since $[y+x\sqrt{A}] = [e][\frac{y}{e}+\frac{x}{e}\sqrt{A}]$, it will suffice to find the canonical form of $[y+x\sqrt{A}]$ in the case in which x and y are relatively prime. In this case, $[y+x\sqrt{A}] = [f, g+\sqrt{A}]$, where $f = |y^2 - Ax^2|$ and g is the least solution mod f of $y \equiv gx \bmod f$, as can be seen in the following way. As was shown in this chapter, the number f defined in this way can be annexed to the list, so the given module is equal to $[f, y+x\sqrt{A}] = [f, y+x\sqrt{A}, ry+rx\sqrt{A}]$ for any r. When r is taken to be the reciprocal of x mod f (which exists, because a common divisor of $f = |y^2 - Ax^2|$ and x is a common divisor of x and y^2), it follows that the given module is equal to $[f, y+x\sqrt{A}, ry+\sqrt{A}] = [f, qf+y-xry, ry+\sqrt{A}]$ for sufficiently large q. Clearly $qf+y-xry \equiv 0 \bmod f$, so the given module is $[f, ry+\sqrt{A}]$

and what is to be shown is that ry satisfies $y \equiv (ry)x \bmod f$, which is clear, and that $(ry)^2 \equiv A \bmod f$, which follows when $y^2 \equiv Ax^2 \bmod f$ is multiplied by r^2.

3. f must be a number modulo which A is a square.

Answers to Exercises for Chapter 19

1. Because 21 is divisible by no square greater than 1, only primitive solutions are possible. There are 4 square roots of 79 mod 21, which can be found by combining the square roots ± 1 of 79 mod 3 and the square roots ± 3 of 79 mod 7 using the Chinese remainder theorem. They are 4, 10, 11, 17. Application of the comparison algorithm to $[21, 4+\sqrt{79}]$ repeats the module $[21, 4+\sqrt{79}]$ on the 8th step without reaching $[1]$, so $g = 4$ corresponds to no solutions of $79\square + 21 = \square$. Its application to $[21, 10+\sqrt{79}]$ reaches $[1]$ on steps 2, 4, 6, ..., which gives the sequence of hypernumbers $(89 + 10\sqrt{79})(80 + 9\sqrt{79})^n$ for $n = 0, 1, 2, \ldots$ whose coefficients solve $79\square + 21 = \square$. Similarly, $[21, 11 + \sqrt{79}]$ leads to $(10 + \sqrt{79})(80 + 9\sqrt{79})^n$ and $[21, 17 + \sqrt{79}]$ leads to no solutions.

2. When the application of the comparison algorithm is organized in the suggested way, it takes the form

$$\begin{array}{ccccccccccccccccccccc} 0 & & 4 & & 5 & & 7 & & 11 & & 13 & & 13 & & 11 & & 7 & & 5 & & 4 \\ & 1 & & 3 & & 4 & & 9 & & 12 & & 13 & & 12 & & 9 & & 4 & & 3 & & 1 \end{array}$$

The numerator of the hypernumber that gives the smallest solution is $(4 + \sqrt{13})^2(5 + \sqrt{13})^2(7 + \sqrt{13})^2(11 + \sqrt{13})^2(13 + \sqrt{13})^2 = (33+9\sqrt{13})^2(90+18\sqrt{13})^2(182+26\sqrt{13}) = 3^2 \cdot 18^2 \cdot 26(11+3\sqrt{13})^2(5+\sqrt{13})^2(7+\sqrt{13})^2$. The numbers in front cancel part of the denominator leaving $(94+26\sqrt{13})^2(7+\sqrt{13}) = 2^2 \cdot 6(83+23\sqrt{13})(47+13\sqrt{13}) = 2^2 \cdot 6 \cdot 12(649 + 180\sqrt{13})$ in the numerator. All the number factors cancel and the solutions of $13\square + 1 = \square$ are the coefficients of $(649 + 180\sqrt{13})^n$ for $n = 0, 1, 2, \ldots$.

4. Eighteen applications of the comparison algorithm to $[1]$ end with $[61, \sqrt{61}]$, after which 18 more steps are needed to return to $[1]$. The hypernumber that describes the smallest nontrivial solution of $61\square+1 = \square$ therefore has in its numerator the square of the product of 18 factors $r_i+\sqrt{61}$ in which the r_i are 8, 10, 16, 14, 13, 11, 9, 11, 19, 21,

17, 19, 31, 41, 49, 55, 59, 61. When these factors are paired and number factors are removed, one finds a product of 9 factors $47 + 6\sqrt{61}$, $19 + 2\sqrt{61}$, $17 + 2\sqrt{61}$, $8 + \sqrt{61}$, $23 + 2\sqrt{61}$, $32 + 3\sqrt{61}$, $37 + 2\sqrt{61}$, $53+2\sqrt{61}$, and $61+2\sqrt{61}$. In a similar way, let the first 8 of these factors be paired and let number factors be removed to find a product of 4 factors $125+16\sqrt{61}$, $58+7\sqrt{61}$, $258+33\sqrt{61}$, and $49+4\sqrt{61}$. Again, this is a number times $(7162 + 917\sqrt{61})(182 + 23\sqrt{61})$, and finally a number times $172669 + 22108\sqrt{61}$. Multiplication by the last of the 9 factors $61 + 2\sqrt{61}$ gives a number times $57(232105 + 29718\sqrt{61})$. In fact, if one keeps track of the omitted number factors, one finds that all are canceled by the denominator and that the required hypernumber is the square of $232105 + 29718\sqrt{61}$ divided by 61. Since the hypernumber that is being squared is $\sqrt{61}(29718+3805\sqrt{61})$, the required hypernumber is the square of $29718+3805\sqrt{61}$. In short, the smallest number x for which $61x^2+1$ is a square is $x = 2{\cdot}3805{\cdot}29718 = 226153980$. This exact answer was given by Bhāskara Achārya in the 12th century.

5. The additional factors in the numerator and denominator of (4) are simply the factors that appear when (4) is used to solve Pell's equation.

6. See, for example, [**E2** p. 31].

Answers to Exercises for Chapter 20

2. Let $Q = (s^2 - A)(Y^2 - AX^2)$ where, by assumption, $s^2 > A$ and $Y^2 > AX^2$. Then $A(Y^2 - AX^2) + Q = s^2(Y^2 - AX^2)$ so $AY^2 + s^2AX^2+Q = s^2Y^2+A^2X^2$. Subtract $2AsXY$ from both sides and use the identity $a^2+b^2-2ab = |a-b|^2$ to find $A|Y-sX|^2+Q = |sY-AX|^2$, which is the required identity when $Y \geq sX$ and $sY \geq AX$.

4. The square of $\frac{1+\sqrt{A}}{2}$ is $\frac{1+2\sqrt{A}+A}{4} = \frac{1+\sqrt{A}}{2} + \frac{A-1}{4}$, so $\frac{1+\sqrt{A}}{2}$ is a root of $X^2 - X - \frac{A-1}{4}$.

5. By elementary algebra, if $y + x\sqrt{A}$ is a root of $X^2 + aX + b$, then $a = -2y$ and $b = y^2 - Ax^2$. Therefore, $y + x\sqrt{A}$ determines a and b and, in particular, determines whether they are integers. (Note that this also proves that $2y$ is an integer. Combined with the fact

that a product of algebraic integers is an algebraic integer, it implies that $2x$ and $2y$ must be integers whenever $y + x\sqrt{A}$ is a unit.)

6. Let $X = y + x\sqrt{A}$ be a root of $X^2 + aX + b$. Division by bX^2 then gives $\frac{1}{b} + \frac{a}{b} \cdot \frac{1}{X} + (\frac{1}{X})^2$. Because $\frac{1}{X}$ is an algebraic integer, Exercise 5 implies that $\frac{1}{b}$ is an integer (as is $\frac{a}{b}$), so $b = \pm 1$.

7. When A is negative, $y^2 - Ax^2$ is a sum of two terms; if it is 1, then, since the second term is at least $|A| \cdot \frac{1}{4}$ unless it is zero, the second term must be zero except in the cases $A = -1$, -2, -3, or -4. Since $x = \frac{1}{2}$ is impossible in the cases $A = -2$ and -4 (in fact, -4 does not arise at all because it is divisible by the square 4), *except when* $A = -1$ *or* -3, *the only units in* $\mathbf{Q}(\sqrt{A})$ *are* ± 1, *so* ± 1 *are fundamental units.* In the remaining two cases, it is easy to show that $\pm\sqrt{-1}$ *are fundamental units in the case* $A = -1$ *and* $\pm\frac{1\pm\sqrt{-3}}{2}$ *are fundamental units in the case* $A = -3$.

8. The equation $Ay^2 + A = z^2$ has a solution if and only if repeated application of the comparison algorithm to $[A, \sqrt{A}]$ reaches $[1]$. (Since A is divisible by no square greater than 1, 0 is the only square root of A mod A that is less than A.) It is first to be shown that $[1]$ is reached if and only if the cycle of $[1]$ is as described. Certainly if the cycle is as described then $[1]$ is reached, because the second half of the steps go from $[A, \sqrt{A}]$ to $[1]$. Suppose now that repeated application of the comparison to $[A, \sqrt{A}]$ does reach $[1]$, say it reaches $[1]$ for the first time on the nth step. What is to be shown is that n more steps return to $[A, \sqrt{A}]$ in a way that mirrors the first n steps—specifically, $r_{n+i} = r_{n+1-i}$ and $f_{n+i} = f_{n-i}$ for $i = 1, 2, \ldots, n$. The lemma of Chapter 26 below shows that the cycle of $[1]$ always shows a symmetry of this type, whether or not it contains $[A, \sqrt{A}]$, so no proof will be given here, but the proof in this specific case is not at all difficult. The formula for the solution of Pell's equation for an A for which the cycle of $[1]$ contains $[A, \sqrt{A}]$ therefore takes the form

$$\frac{\left((r_1 + \sqrt{A})(r_2 + \sqrt{A}) \cdots (r_n + \sqrt{A})\right)^2}{\left(f_1 f_2 \cdots f_{n-1}\right)^2 \cdot A}$$

where $r_1, r_2, \ldots, r_n$ and $f_1, f_2, \ldots, f_{n-1}$ are the numbers (in reverse order) used to construct the hypernumber $z_2 + y_2\sqrt{A}$ that gives the smallest solution of $z_2^2 = Ay_2^2 + A$. The assumption that A has no square divisors greater than 1 easily implies that $z_2 \equiv 0 \bmod A$ and the remaining statements follow.

9. In the first case, when $-1 = y^2 - Ax^2$ has no solution in numbers, the coefficients of a unit must, by Exercise 6, satisfy $1 = y^2 - Ax^2$. Thus, if x and y are integers, $|x|$ and $|y|$ must be numbers that satisfy Pell's equation. Therefore $|y|+|x|\sqrt{A} = (y_1+x_1\sqrt{A})^n$ for some number n. Since $(y_1 + x_1\sqrt{A})(y_1 - x_1\sqrt{A}) = 1$, it follows that $|y| - |x|\sqrt{A} = (y_1 + x_1\sqrt{A})^{-n}$, so $\pm(|y| \pm |x|\sqrt{A}) = \pm(y_1 + x_1\sqrt{A})^{\pm n}$ and all four possible sign combinations are taken on by the formula $\pm\varepsilon^n$. The second case is similar.

10. If there are such units, there are odd numbers x and y for which $y^2 - Ax^2 = \pm 4$. Since the square of an odd number is 1 mod 8, it follows that $1 \equiv A \pm 4 \bmod 8$, so $A \equiv 5 \bmod 8$.

11. If the norm of $\frac{y+x\sqrt{A}}{2}$ is -1 then the norm of its square is 1 and its square $\frac{y^2+Ax^2+2xy\sqrt{A}}{4}$ is $\frac{y'+x'\sqrt{A}}{2}$ where x' and y' are odd. (In fact, $y' \equiv 3 \bmod 4$.) Therefore, such a unit implies a primitive solution of $y^2 = Ax^2 + 4$, which implies that repeated application of the comparison algorithm to $[4, g+\sqrt{A}]$ reaches $[1]$ for the square root g of 3 mod 4 that is determined by $y \equiv gx \bmod 4$. This is the needed condition. The methods of Chapter 26 easily show that this is true if and only if $[4, 1+\sqrt{A}]$ or $[4, 3+\sqrt{A}]$ is in the cycle of $[1]$, and in fact show that it is true if and only if *both* of them are in the cycle of $[1]$. This is not the case when $A = 37$.

12. When $A = 21$ the successors of $[1]$ are $[4, 1]$, $[7, 0]$, $[4, 3]$, and $[1]$. The last step uses $r = 5$ and corresponds to the equation $21 \cdot 1^2 + 4 = 5^2$, which leads to the unit $\frac{5+\sqrt{21}}{2}$ whose square is $\frac{23+5\sqrt{21}}{2}$ and whose cube is $55 + 12\sqrt{21}$, which gives the smallest solution of Pell's equation. The method is clearer in the less trivial example $A = 69$. The successors $[f_1, g_1 + \sqrt{69}]$, $[f_2, g_2 + \sqrt{69}]$, ... of $[1]$ return to $[1]$ for the first time at $[f_8, g_8 + \sqrt{69}] = [1]$. Because $[f_5, g_5 + \sqrt{69}] = [4, 1 + \sqrt{69}]$ the smallest solution of $Ax^2 + 4 = y^2$ is given by

$$25 + 3\sqrt{69} = \frac{(11 + \sqrt{69})(15 + \sqrt{69})(9 + \sqrt{69})}{13 \cdot 12},$$

so $\frac{25+3\sqrt{69}}{2}$ is a unit. Its square is the unit $\frac{623+75\sqrt{69}}{2}$ (the equation $69 \cdot 75^2 + 4 = 623^2$ can also be found by applying the theorem of Chapter 19 beginning with the module $[f_3, g_3 + \sqrt{69}] = [4, 3]$) and its cube is $7775 + 936\sqrt{69}$, whose coefficients are the smallest solution of Pell's equation. Therefore, $\frac{25+3\sqrt{69}}{2}$ is a fundamental unit. In the same way, whenever the cycle of [1] contains $[4, 1 + \sqrt{A}]$ but not $[A, \sqrt{A}]$ (the next case is 77) one can find a fundamental unit using the steps of the comparison algorithm from the last of the two $[4, 1 + \sqrt{A}]$ and $[4, 3 + \sqrt{A}]$ to occur in the cycle up to the first occurrence of [1] to find a solution of $Ax^2 + 4 = y^2$ and derive from it a fundamental unit.

13. $A = 5$. By the comparison algorithm $[5 + \sqrt{5}][20, 15 + \sqrt{5}] = [20]$, so multiplication by $[20, 5 + \sqrt{5}]$ and division by [20] gives $[5 + \sqrt{5}] = [20, 5 + \sqrt{5}]$. Thus $5^2 - 5 \cdot 1^2 = 20$ and division by 20 gives $5 \cdot (\frac{1}{2})^2 - (\frac{1}{2})^2 = 1$. Thus, $\frac{1+\sqrt{5}}{2}$ is a unit. Its cube is $2 + \sqrt{5}$, which gives the smallest solution $(x, y) = (1, 2)$ of $y^2 + 1 = 5x^2$, and the fundamental unit is $\frac{1+\sqrt{5}}{2}$.

$A = 13$. By the comparison algorithm, $[5 + \sqrt{13}][19 + \sqrt{13}][39 + \sqrt{13}][52, 13 + \sqrt{13}] = [12][29][52]$, from which $[13 + 3\sqrt{13}] = [52, 39 + \sqrt{13}]$. It follows that $\frac{3+\sqrt{13}}{2}$ has norm -1. Its cube is $18 + 5\sqrt{13}$ which gives the smallest solution of $y^2 + 1 = 13x^2$, and a fundamental unit is $\frac{3+\sqrt{13}}{2}$.

$A = 29$. Here $[116, 29 + \sqrt{29}] = [29 + 5\sqrt{29}]$ leads to the fundamental unit $\frac{5+\sqrt{29}}{2}$.

$A = 53$. Application of the comparison algorithm to $[212, 53 + \sqrt{53}]$ leads to $[212, 159 + \sqrt{53}] = [53 + 7\sqrt{53}]$. A fundamental unit is $\frac{7+\sqrt{53}}{2}$.

$A = 61$. In this case, $[244, 183 + \sqrt{61}] = [305 + 39\sqrt{61}]$ and a fundamental unit is $\frac{39+5\sqrt{61}}{2}$. (The square of this unit is the unit $\frac{1523+195\sqrt{61}}{2}$ and its cube is the unit $29718 + 3805\sqrt{61}$ with norm -1. The square of this unit is the sixth power of the fundamental unit and the smallest solution of Pell's equation. Thus, the smallest x for which $61x^2 + 1$ is a square is $x = 2 \cdot 29718 \cdot 3805$, as was found in Exercise 4 of Chapter 19.)

$A = 85$. Here $[340, 85 + \sqrt{85}] = [85 + 9\sqrt{85}]$ and a fundamental unit is $\frac{9+\sqrt{85}}{2}$.

$A = 109$. $[436, 109 + \sqrt{109}] = [2725 + 261\sqrt{109}]$. Fundamental unit is $\frac{261+25\sqrt{109}}{2}$. (The cube of the fundamental unit is $8890182 + 851525\sqrt{109}$ and the smallest x for which $109x^2 + 1$ is a square is the prodigious number $2 \cdot 8890182 \cdot 851525 = 15140424455100$.)

Answers to Exercises for Chapter 21

1. When $A = 2$, circles go around 1 and 7, squares around 3 and 5. For $A = 3$ the circles are 1 and 11, the squares 5 and 7. For $A = 5$, circles are 1, 9, 11, 19 and squares are 3, 7, 13, 17. For $A = 6$, circles are 1, 5, 19, 23 and squares are 7, 11, 13, 17. In all cases, if k gets a circle, then so does $4A - k$.

2. $\phi(60) = 16$. The squares 1 and 49 and their negatives 11 and 59 account for 4 circles. Since 15 is a square mod 7, 7 times any circled number is circled, which accounts for another 4 and there are no others.

3. If A is an odd square, the squares of $1, 3, 5, \ldots, A-2$ are distinct mod A (because no two of these numbers sum to 0 mod A) and there are $\frac{A-1}{2}$ of them, which gives $\frac{A-1}{2}$ numbers less than $4A$ that must be circled. Subtracting one of these from $4A$ gives another number that must be circled, which gives $\frac{A-1}{2}$ new numbers that must be circled, because they are distinct from each other and distinct from the first set of $\frac{A-1}{2}$ because they are all 3 mod 4 and the first set, being squares of odd numbers, are all 1 mod 4. Thus, the $A - 1$ circled numbers are all accounted for. Conclusion: Let A and p be distinct odd prime numbers. If $p \equiv 1 \bmod 4$, then $A \equiv \square \bmod p$ if and only if $p \equiv \square \bmod A$, but if $p \equiv 3 \bmod 4$, then $A \equiv \square \bmod p$ if and only if $-p \equiv \square \bmod A$. (This is the way Gauss states the "fundamental theorem" in Article 131 of [**G**].)

Answers to Exercise for Chapter 23

1. (a) The content is $[3, 0, \frac{75}{3}] = [3, 25] = [1]$. (b) The square is $[9, 3\sqrt{75}, 75] = [3, 3\sqrt{75}] = [3] \sim [1]$, so $[3, \sqrt{75}]$ is primitive by

the proposition of this chapter. (c) The successor is $[2, 1 + \sqrt{75}]$, its product with $[3, \sqrt{75}]$ is $[6, 3 + \sqrt{75}]$ by the Chinese remainder theorem, and the successor of $[6, 3 + \sqrt{75}]$ is $[1]$, so again the module is primitive by virtue of the proposition.

Answers to Exercises for Chapter 24

2. When f and F are relatively prime, $[f, g + \sqrt{A}][F, G + \sqrt{A}]$ has the form $[fF, \mathcal{G} + \sqrt{A}]$ for some $\mathcal{G}$, in which case the signature of the product is clearly the product of the signatures. More interesting examples are $[2, \sqrt{10}][2, \sqrt{10}] = [2]$, where both factors have signature $--$ and the product has signature $++$, and $[15, \sqrt{15}][3, \sqrt{15}] = [3][5, \sqrt{15}] \sim [2, 1 + \sqrt{15}]$, where the factors have signatures $-+-$ and $+--$ and the product has signature $--+$. Note that the rule often *fails* when one or both factors are not primitive.

Answers to Exercises for Chapter 26

1. The first examples are $[7, 3 + \sqrt{37}]$ and $[7, 4 + \sqrt{37}]$. Others occur for $A = 79$, 99, and 101.

2. See Exercise 1.

Answers to Exercises for Chapter 27

1. In the range of the table in the appendix, only $A = 37$ and 101 have the desired property. Extend the table to find others.

2. In the range of the table, the only example is $A = 79$. Extend the table to find others.

Answers to Exercises for Chapter 30

1. Set $\begin{bmatrix} q & r \\ s & t \end{bmatrix} = \begin{bmatrix} 1 & n \\ 0 & 1 \end{bmatrix}$ in formula (2) to find that $ax^2 + 2(b + na)xy + (c + 2nb + n^2a)y^2$ is equivalent to $ax^2 + 2bxy + cy^2$. For sufficiently large n, $b + na$ is positive (as is $c + 2nb + n^2a$).

2. The change of variables $x = x_1 + 2y_1$, $y = y_1$ transforms $2x^2 - 8xy + 3y^2$ to $2x_1^2 - 5y_1^2$ and $x_1 = Y$, $y_1 = -X$ transforms $2x_1^2 - 5y_1^2$ to $-5X^2 + 2Y^2$. What is needed, therefore, is a transformation of $5X^2 - 2Y^2$ to $13u^2 + 12uv + 2v^2$. These correspond to $[5, \sqrt{10}]$ and $[13, 6 + \sqrt{10}]$, respectively. The first is stable and application of the comparison algorithm to $[13, 6+\sqrt{10}]$ gives $[15+4\sqrt{10}][13, 6+\sqrt{10}] = [13][5, \sqrt{10}]$ in two steps. Formula (3) then gives $u = 3X + 2Y$ and $v = 4X + 3Y$ for the desired transformation. Composed with $X = -y_1 = -y$ and $Y = x_1 = x - 2y$, this transformation is $u = 2x - y$, $v = 3x - 2y$, the very transformation given by Gauss.

3. The equation

$$\begin{bmatrix} p & 0 \\ 0 & -1 \end{bmatrix} = \begin{bmatrix} 0 & 1 \\ -1 & 0 \end{bmatrix} \begin{bmatrix} -1 & 0 \\ 0 & p \end{bmatrix} \begin{bmatrix} 0 & -1 \\ 1 & 0 \end{bmatrix}$$

shows that $-u^2 + pv^2$ is equivalent to $px^2 - y^2$, so the question is whether the module $[p, \sqrt{p}]$ is equivalent to $[1, \sqrt{p}]$, or, more simply, whether $[p, \sqrt{p}]$ is in the principal cycle. As was seen in Chapter 27, this is the case if and only if $p \equiv 1 \bmod 4$ or $p = 2$.

4. The proof is exactly the same as in Chapter 17, except that the possibility of subtraction allows for some slight simplifications.

5. The identity $r^2 + (g+\sqrt{-A})(r+\sqrt{-A}) = \frac{r+g}{f} \cdot f \cdot (r+\sqrt{-A}) - A$ analogous to (2) of Chapter 19 implies $[r + \sqrt{-A}][f, g + \sqrt{-A}] = [f(r + \sqrt{-A}), (g + \sqrt{-A})(r + \sqrt{-A})] = [f(r + \sqrt{-A}), (g + \sqrt{-A})(r + \sqrt{-A}), r^2 + A] = [f(r + \sqrt{-A}), (g + \sqrt{-A})(r + \sqrt{-A}), fF] = [f(r + \sqrt{-A}), fF]$, from which $[f, g+\sqrt{-A}] \sim [F, G+\sqrt{-A}]$ follows, where r is the least nonnegative solution of $r + g \equiv 0 \bmod f$, $F = (r^2 + A)/f$, and G is the least nonnegative solution of $r \equiv G \bmod F$.

6. By Euler's criterion $C_p(p-1) \equiv (p-1)^{(p-1)/2} \equiv 1 \bmod p$, so $p - 1 \equiv g^2 \bmod p$ for some positive g. For example, when $p = 29$, $[29, 17 + \sqrt{-1}]$ is in canonical form. The comparison algorithm as in Exercise 5 gives $[29, 17 + \sqrt{-1}] \sim [5, 2 + \sqrt{-1}] \sim [2, 1 + \sqrt{-1}] \sim [1]$ and, as in Chapter 19, $[29, 17+\sqrt{-1}] = [a+b\sqrt{-1}]$ where $a+b\sqrt{-1} = \frac{(12+\sqrt{-1})(3+\sqrt{-1})(1+\sqrt{-1})}{5\cdot 2} = 2 + 5\sqrt{-1}$, which yields the solution $29 = 2^2 + 5^2$. The method works for every $p \equiv 1 \bmod 4$ because $r < f$ (the least nonnegative solution of $r + g \equiv 0 \bmod f$ is less than f) so

$F = \frac{r^2+1}{f} \le \frac{(f-1)^2+1}{f} = f - 2 + \frac{2}{f}$; thus $F < f$ unless $f = 1$, so f decreases until 1 is reached and the method of Chapter 19 gives a representation $[p, g + \sqrt{-A}] = [a + b\sqrt{-A}]$ and $a^2 + b^2 = p$.

7. In the notation of Chapter 24, $C_p(-2) = C_p(-1)C_p(2) = \lambda_1(p)\lambda_2(p) = \lambda_3(p)$ so $[p, g + \sqrt{-2}]$ is in canonical form for some g if and only if p is 1 or 3 mod 8. That the comparison algorithm must reach [1] (after which it simply alternates between [1] and $[2, \sqrt{-2}]$) follows from $F = \frac{r^2+2}{f} \le \frac{(f-1)^2+2}{f} = \frac{f^2-2f+3}{f} = f - \frac{2f-3}{f} < f$ for $f > 1$.

8. $C_p(-3) = \lambda_1(p)C_p(3)$ is 1 if and only if $p \equiv 1 \bmod 3$ as one finds by making use of Euler's law and $C_5(3) = -1$, $C_7(3) = -1$, and $C_{11}(3) = 1$. In this case, the method of Exercise 6 gives $F < f$ whenever $f > 2$ so the comparison algorithm applied to $[p, g + \sqrt{-3}]$ must reach [1] unless it reaches $[2, 1 + \sqrt{-3}]$. But the latter is impossible because the content of $[2, 1 + \sqrt{-3}]$ is 2 and the content of $[p, g + \sqrt{-3}]$ is 1.

9. Application of the comparison algorithm to $[29, 13 + \sqrt{-5}]$ reaches [1], but its application to $[7, 3 + \sqrt{-5}]$ cycles through $[3, 1 + \sqrt{-5}]$, $[3, 2 + \sqrt{-5}]$, and $[2, 1 + \sqrt{-5}]$. There are two equivalence classes of modules in this case, and a module $[f, g + \sqrt{-5}]$ in canonical form in which f is relatively prime to 10 is in the principal class if $f \equiv 1 \bmod 4$ and $f \equiv \square \bmod 5$, in the other class if $f \equiv -1 \bmod 4$ and $f \not\equiv \square \bmod 5$. Thus, $p = \square + 5\square$ if and only if $p \equiv 1$ or $9 \bmod 20$. A product of *two* primes p and q that are both 3 or 7 mod 20 has the form $pq = \square + 5\square$. These facts were known to Fermat, but he confessed he was unable to prove them.

Answers to Exercises for Chapter 31

1. In this case, $[sa, sb + \sqrt{A}] = [1, \sqrt{A}]$ and $\sigma = 1$, so formula (2) implies $E = 1$, $F = \alpha$, and $G = \beta$, from which $Y = \alpha yu + \beta yv + xv$ and then $X = xu - \beta yu - \gamma yv$. When X and Y are defined in this way, the formula $(x^2 - Ay^2)(\alpha u^2 + 2\beta uv + \gamma v^2) = \alpha X^2 + 2\beta XY + \gamma Y^2$ holds.

2.

$$\begin{aligned}[3, 1+\sqrt{-140}][9, 7+\sqrt{-140}] \\ &= [27, 21+3\sqrt{-140}, 9+9\sqrt{-140}, -133 \; + \; 8\sqrt{-140}] \\ &= [27, 21+3\sqrt{-140}, 142+\sqrt{-140}, -133+8\sqrt{-140}] \\ &= [27, 21+3\sqrt{-140}, 7+\sqrt{-140}, -133+8\sqrt{-140}] \\ &= [27, 7+\sqrt{-140}].\end{aligned}$$

In this case $s = \sigma = 1$, $a = 3$, $b = 1$, $\alpha = 9$, $\beta = 7$, $E = 1$, $F = 27$, $G = 7$ from which

$$(3x^2 + 2xy + 47y^2)(9u^2 + 14uv + 21v^2) = 27X^2 + 14XY + 7Y^2$$

where $X = xu - 2yu - 7yv$, $Y = 3xv + 9yu + 8yv$.

3. By assumption, there is a change of variables $x_1 = qx + ry$, $y_1 = sx + ty$ with $qt - rs = 1$ that transforms $a_1x_1^2 + 2b_1x_1y_1 + c_1y_1^2$ to $ax^2 + 2bxy + cy^2$. This same change of variables in X_1 and Y_1 gives polynomials X_2 and Y_2 in x, y, u, and v for which $(ax^2 + 2bxy + cy^2)(\alpha u^2 + 2\beta uv + \gamma v^2) = \mathcal{A}_1X_2^2 + 2\mathcal{B}_1X_2Y_2 + \mathcal{C}_1Y_2^2$. Thus, the two forms $\mathcal{A}X^2 + 2\mathcal{B}XY + \mathcal{C}Y^2$ and $\mathcal{A}_1X^2 + 2\mathcal{B}_1XY + \mathcal{C}_1Y^2$ compose the same two forms so they are equivalent by Gauss's theorem.

Bibliography

[D] L. E. Dickson, *History of the Theory of Numbers,* Carnegie Institute, Washington, 1920, Chelsea reprint, 1971.

[E1] H. M. Edwards, *Riemann's Zeta Function,* Academic Press, New York, 1974, Dover reprint, 2001.

[E2] H. M. Edwards, *Fermat's Last Theorem,* Springer-Verlag, New York, 1974.

[E3] H. M. Edwards, *Essays in Constructive Mathematics,* Springer-Verlag, New York, 2005.

[E4] H. M. Edwards, *Composition of Binary Quadratic Forms and the Foundations of Mathematics,* article in *The Shaping of Arithmetic,* C. Goldstein et al., eds., Springer-Verlag, Berlin, Heidelberg, New York, 2007.

[G] C. F. Gauss, *Disquisitiones Arithmeticae,* Braunschweig, 1801. (Reprinted as vol. 1 of Gauss's *Collected Works* (*Gesammelte Werke*) and available in many editions in which it is translated into many languages.)

[J] C. G. J. Jacobi, *Canon Arithmeticus,* Berlin, Typis Academicis, 1839.

Index

Titles in This Series

TITLES IN THIS SERIES

17 **A. Shen and N. K. Vereshchagin,** Basic set theory, 2002

16 **Wolfgang Kühnel,** Differential geometry: curves – surfaces – manifolds, second edition, 2006

15 **Gerd Fischer,** Plane algebraic curves, 2001

14 **V. A. Vassiliev,** Introduction to topology, 2001

13 **Frederick J. Almgren, Jr.,** Plateau's problem: An invitation to varifold geometry, 2001

12 **W. J. Kaczor and M. T. Nowak,** Problems in mathematical analysis II: Continuity and differentiation, 2001

11 **Michael Mesterton-Gibbons,** An introduction to game-theoretic modelling, 2000

10 **John Oprea,** The mathematics of soap films: Explorations with Maple®, 2000

9 **David E. Blair,** Inversion theory and conformal mapping, 2000

8 **Edward B. Burger,** Exploring the number jungle: A journey into diophantine analysis, 2000

7 **Judy L. Walker,** Codes and curves, 2000

6 **Gérald Tenenbaum and Michel Mendès France,** The prime numbers and their distribution, 2000

5 **Alexander Mehlmann,** The game's afoot! Game theory in myth and paradox, 2000

4 **W. J. Kaczor and M. T. Nowak,** Problems in mathematical analysis I: Real numbers, sequences and series, 2000

3 **Roger Knobel,** An introduction to the mathematical theory of waves, 2000

2 **Gregory F. Lawler and Lester N. Coyle,** Lectures on contemporary probability, 1999

1 **Charles Radin,** Miles of tiles, 1999